This Book belongs to:

_____

- -

_____

Index

This handwriting practice book is designed for children aged 6-8 years who are working on their handwriting skills, learning the alphabet, numbers and high frequency words.

It is organized starting with the alphabet and moves onto numbers, a selection of high frequency sight words and finally number words. We suggest adult guidance initially and you can either work through the entire book in sequence or go straight to sections of most need.

Section 1: Tracing the alphabet both upper and lowercase

Section 2: Writing numbers 1-100

Section 3: Selection of high frequency words

Section 4: Writing number words 1-100

Throughout the book we have blank lines to practice unaided writing and included coloring pictures to provide regular fun breaks from the handwriting practice.

The line spacing in sections 3 & 4 is reduced as a progression from sections 1 & 2.

Section 1:
Learning the Alphabet

Trace the letters, both upper and lowercase, and practice writing them on your own on the blank lines provided.

Color in the Zebra

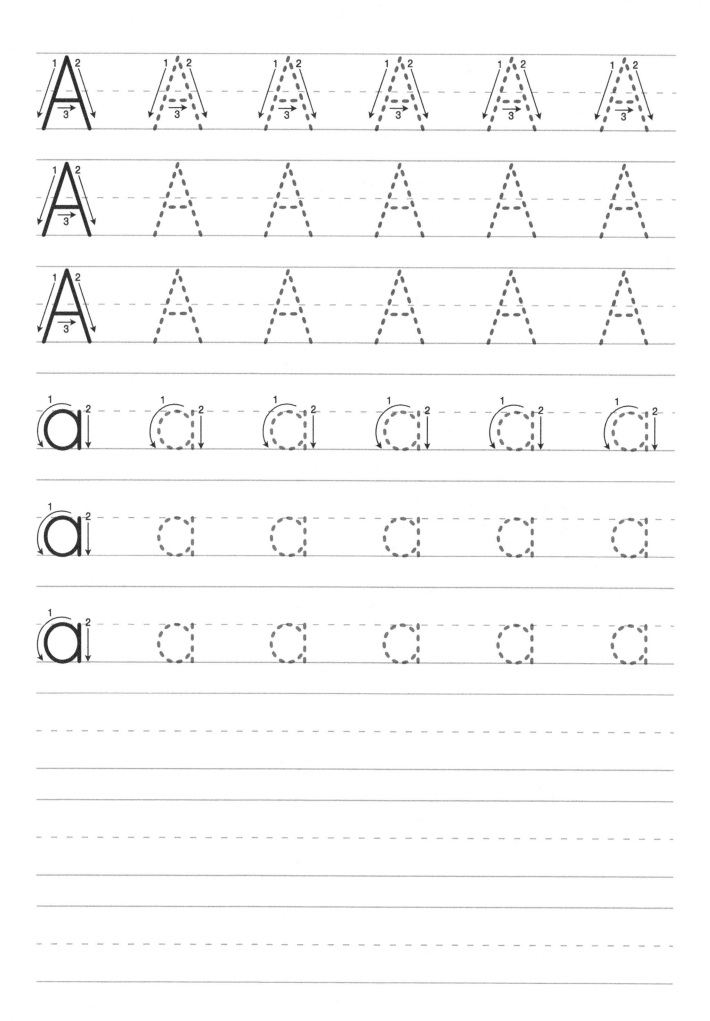

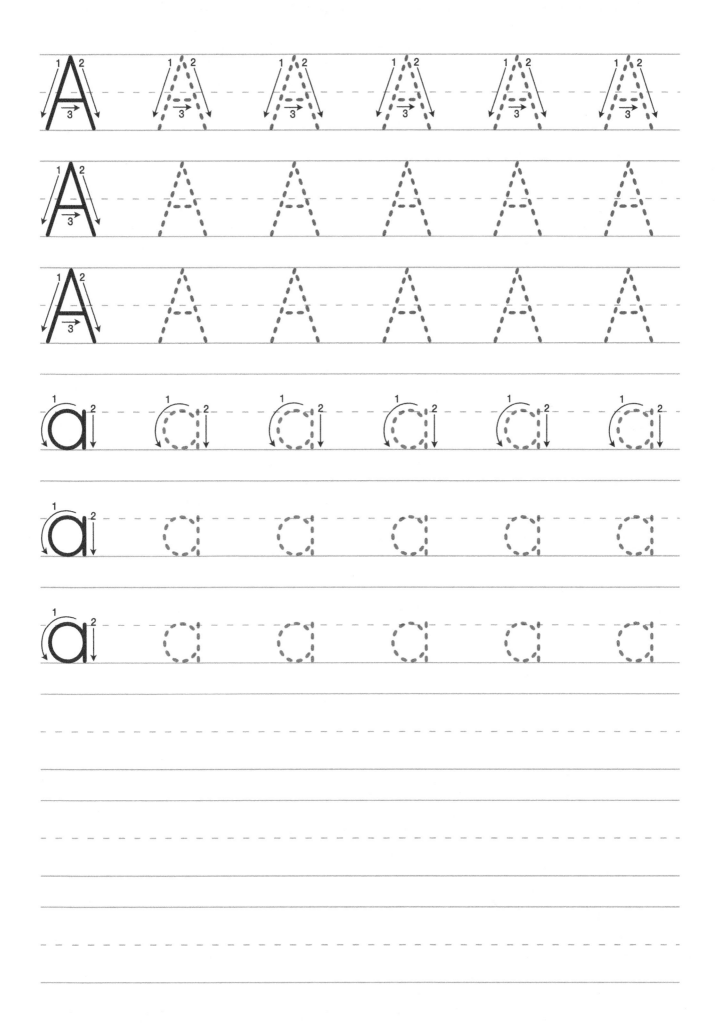

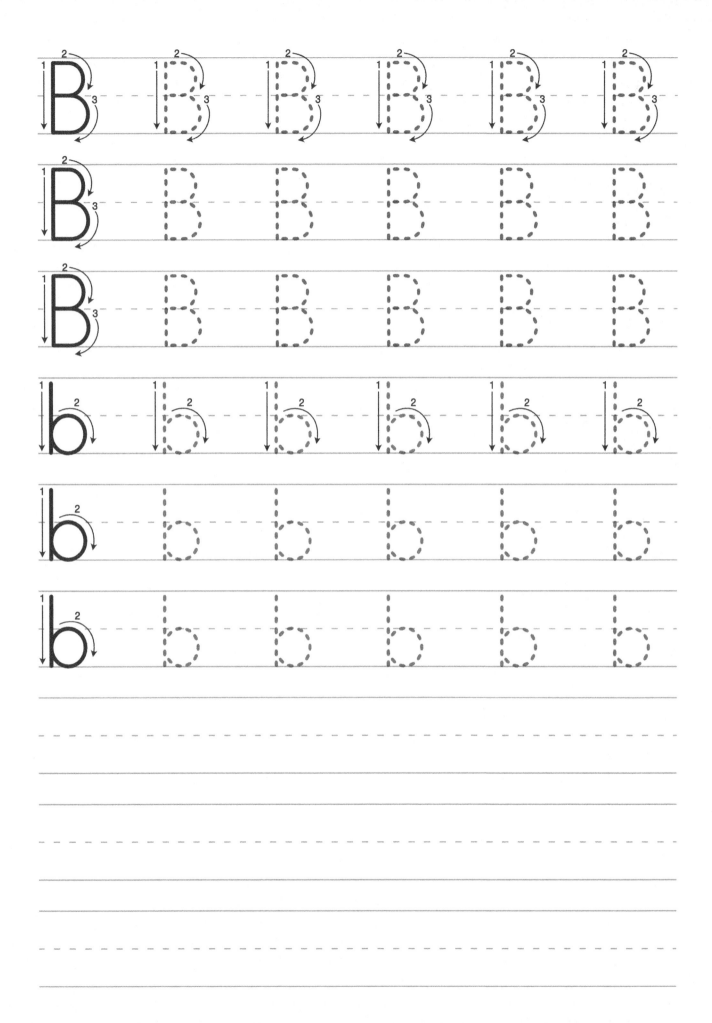

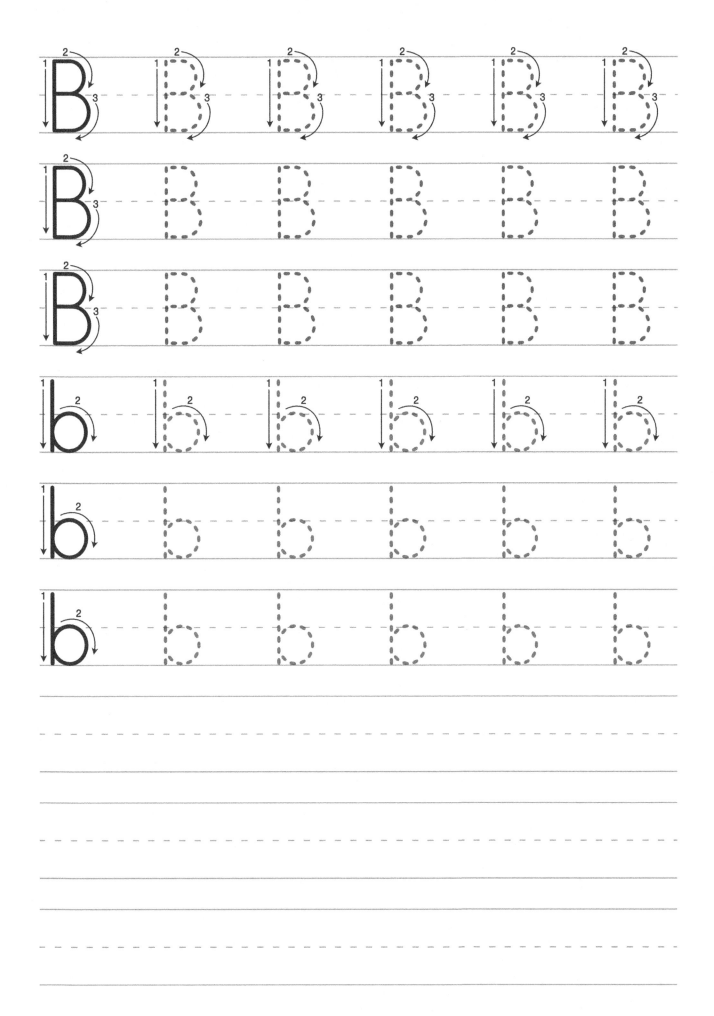

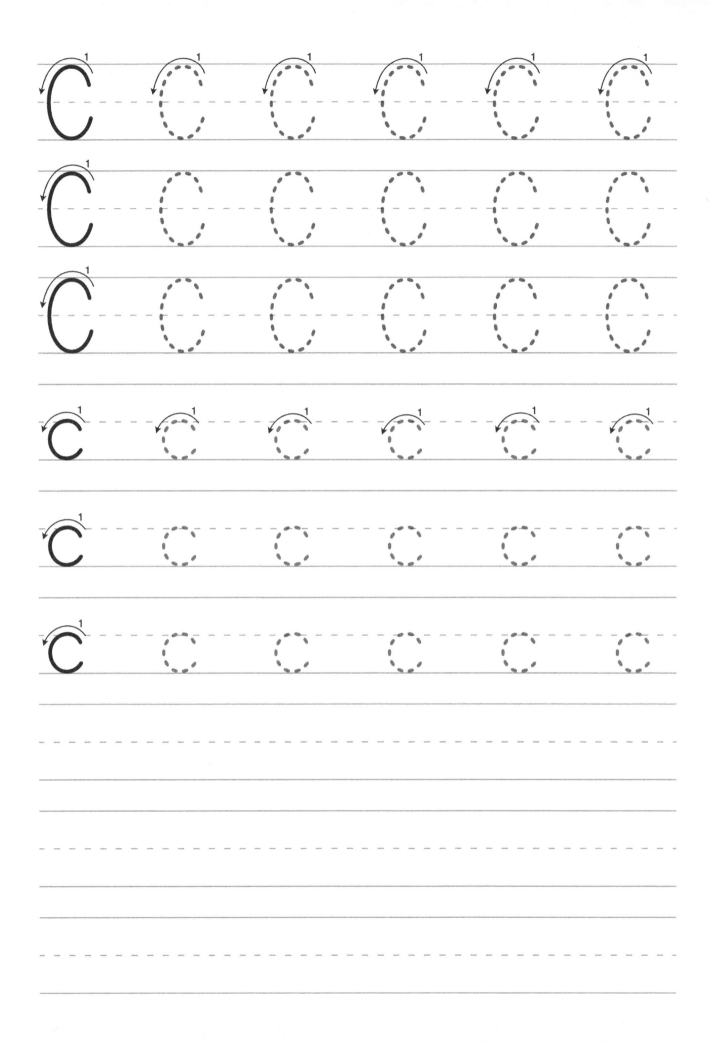

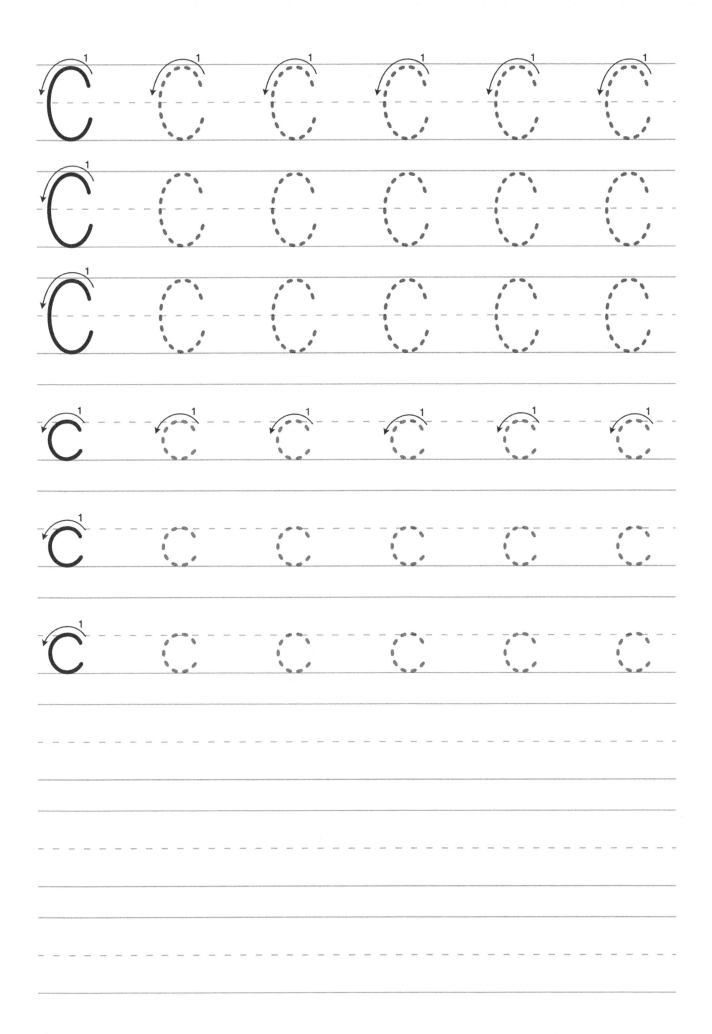

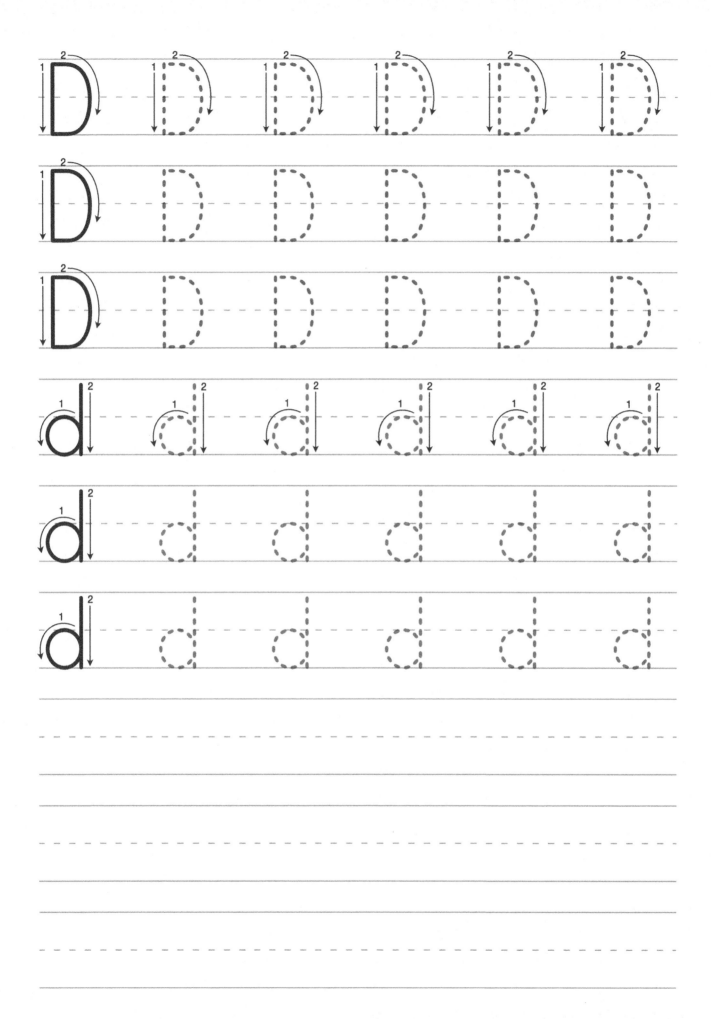

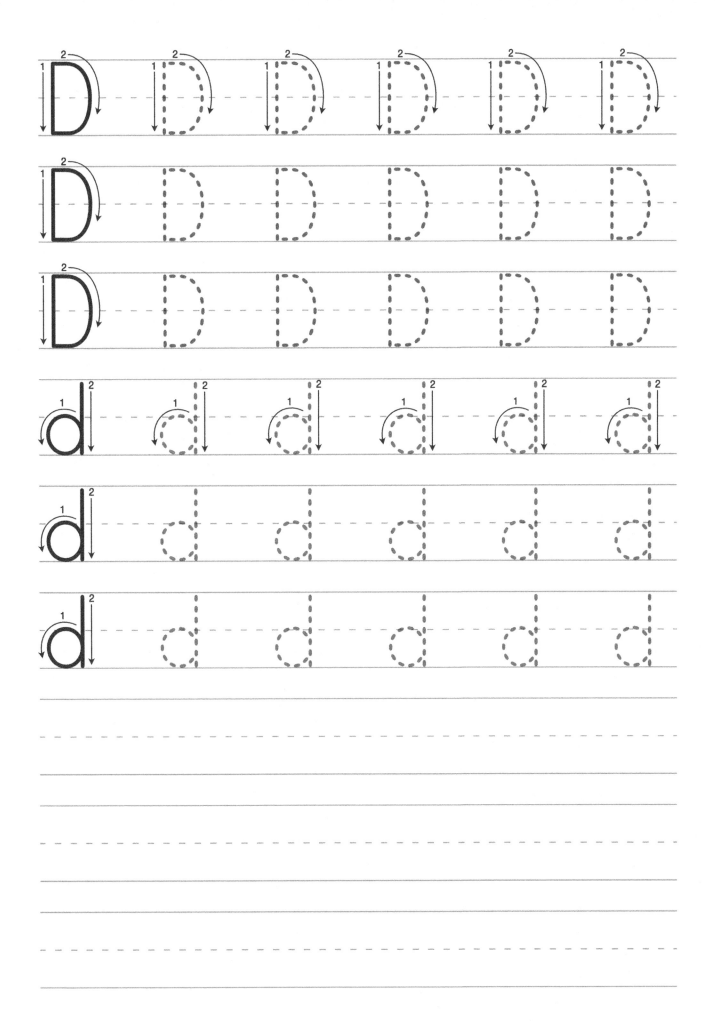

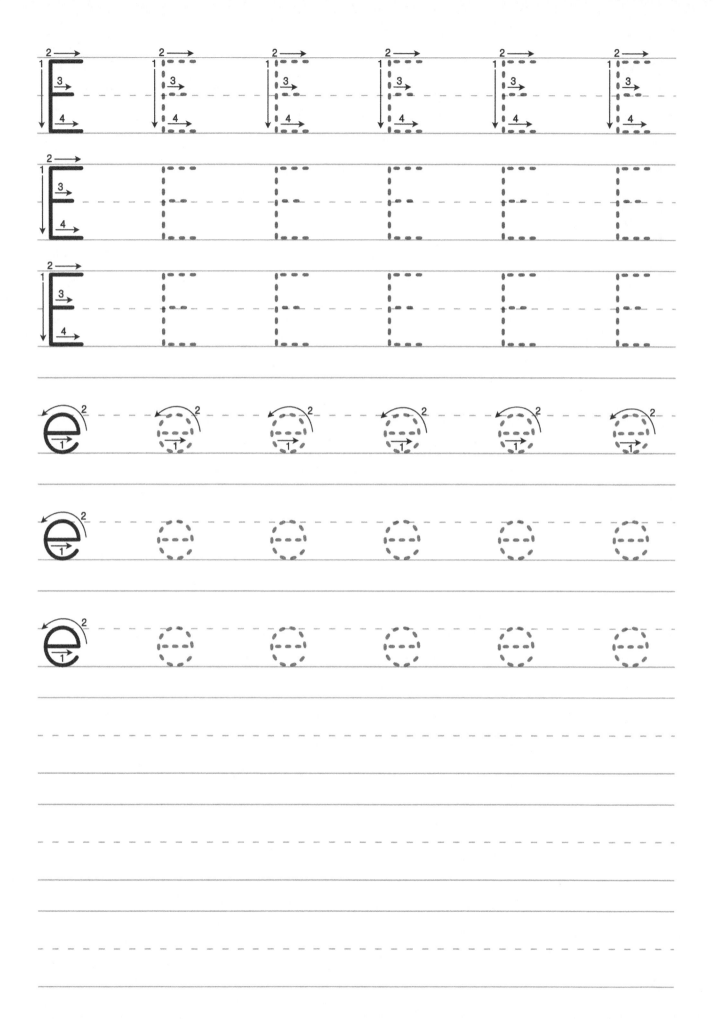

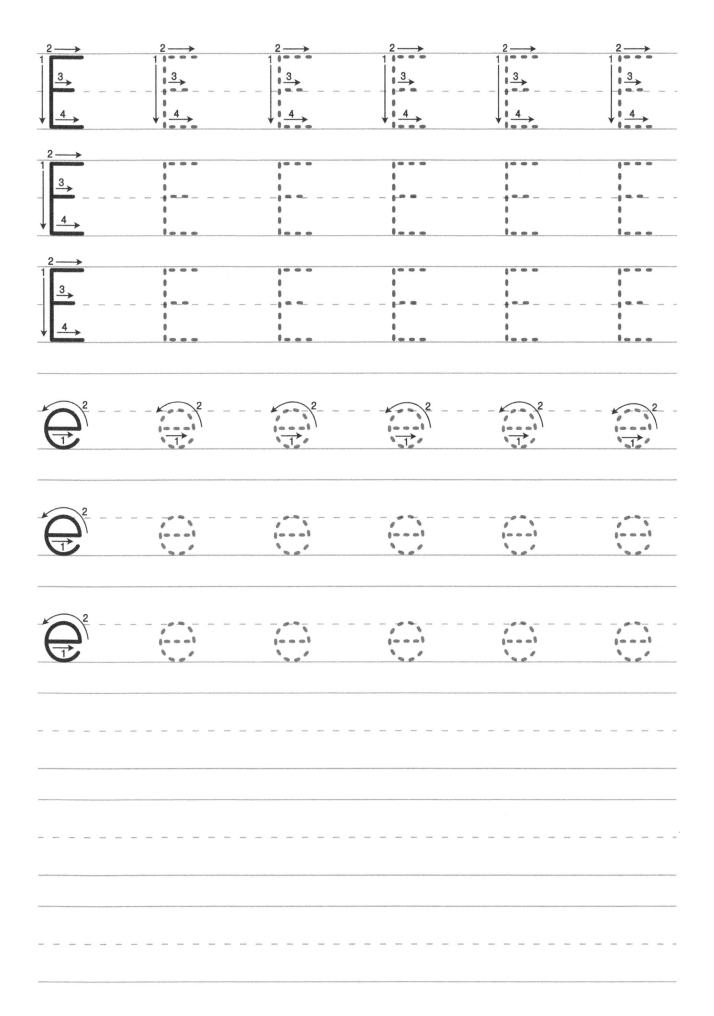

Use the space below to practice your writing.

Color in the Tiger

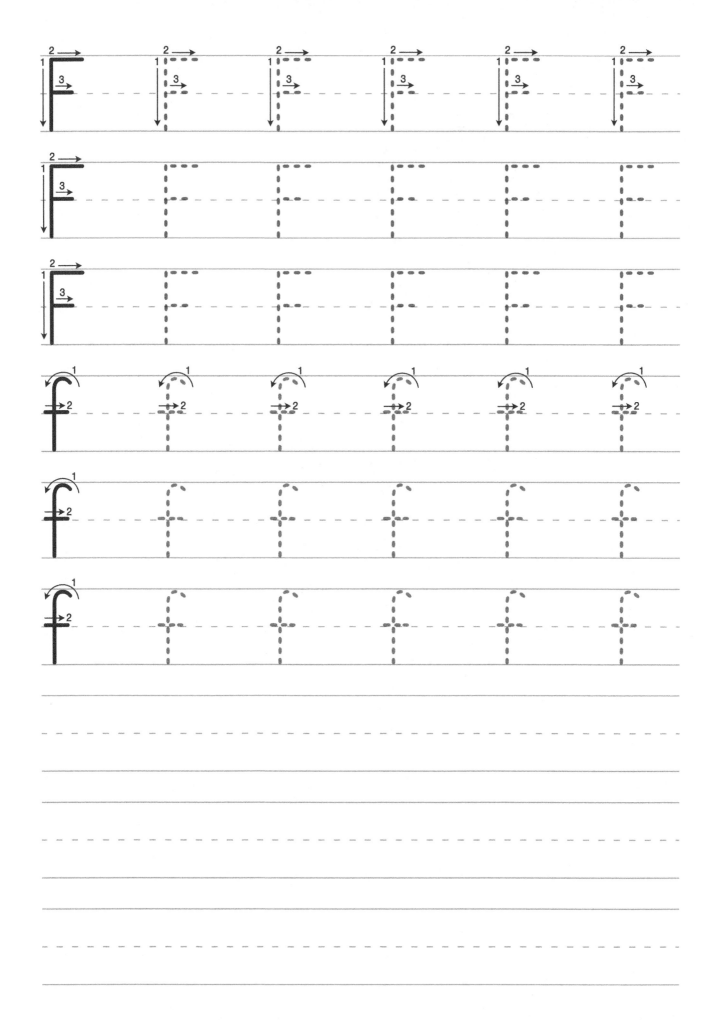

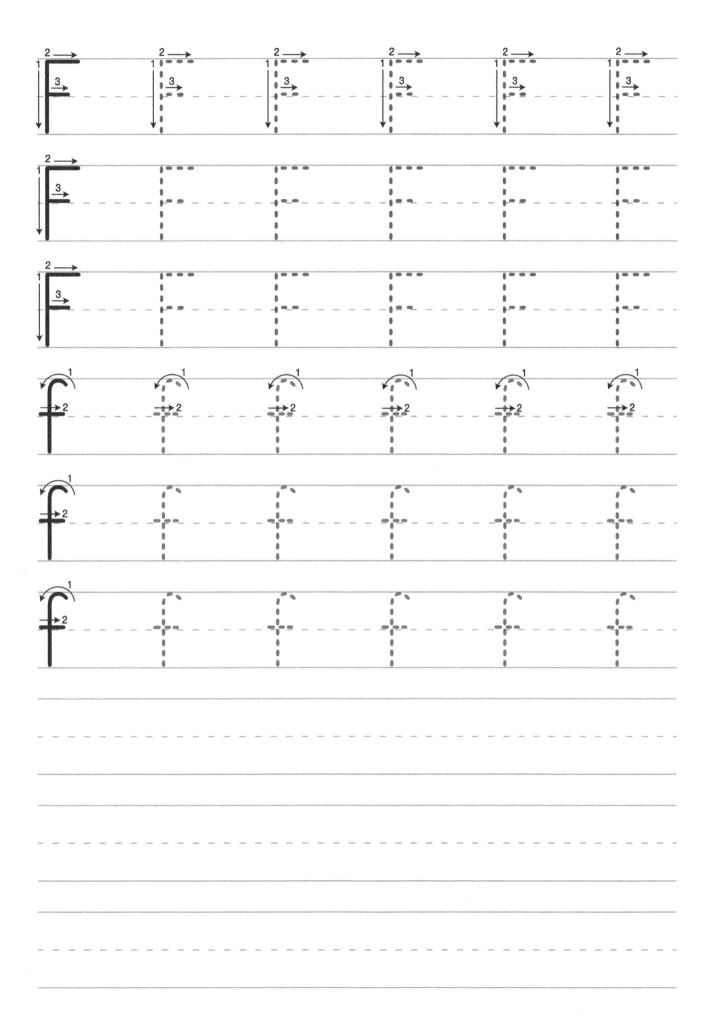

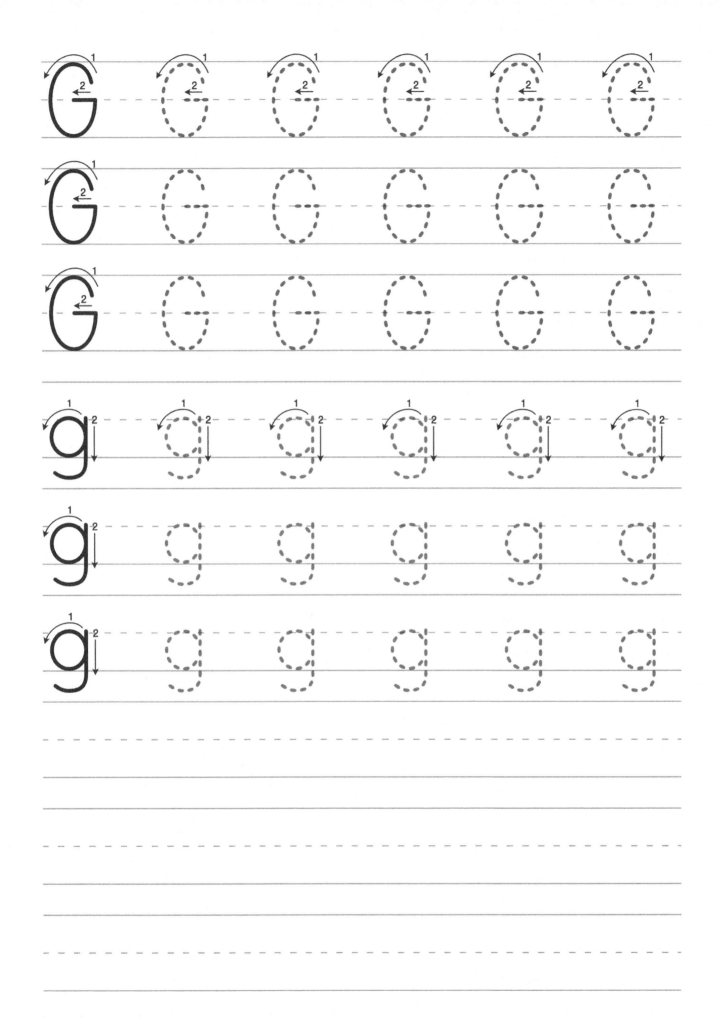

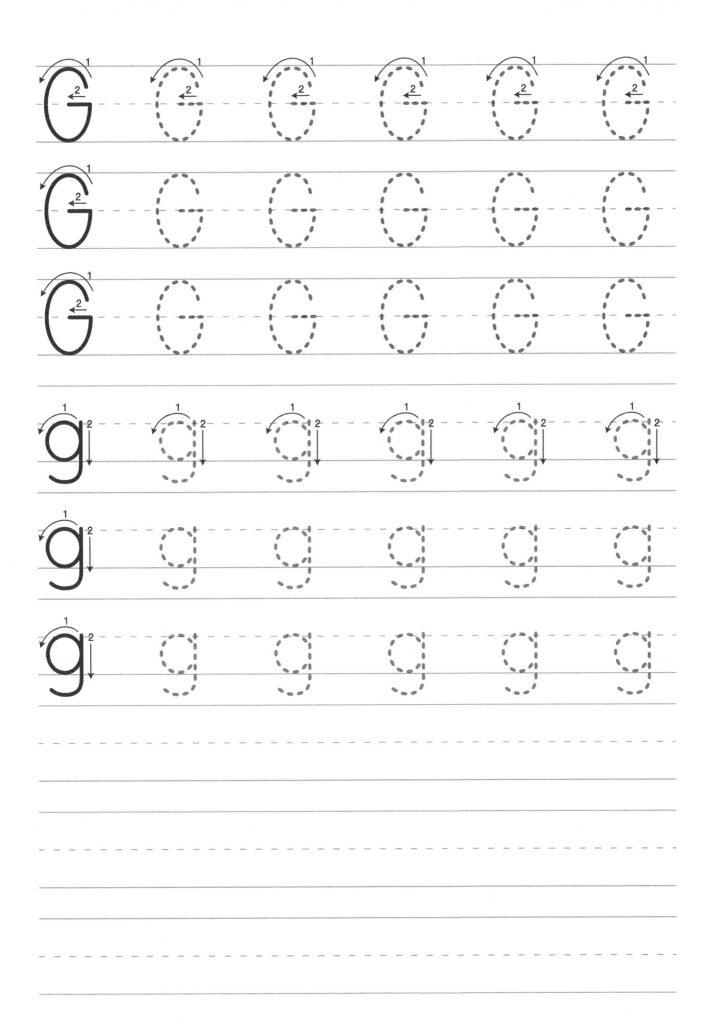

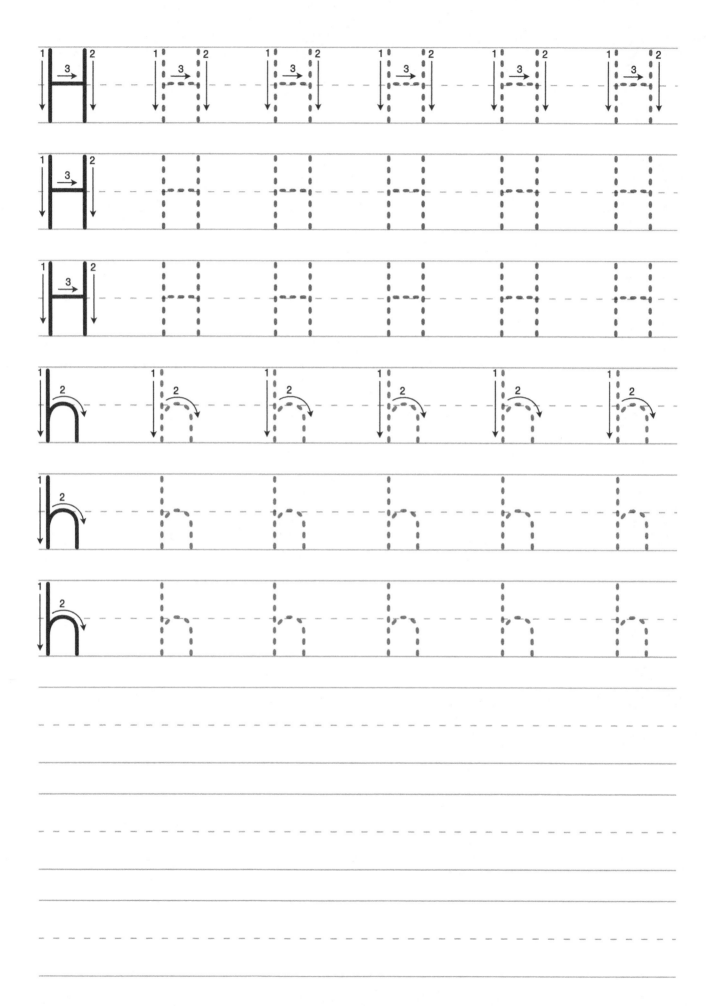

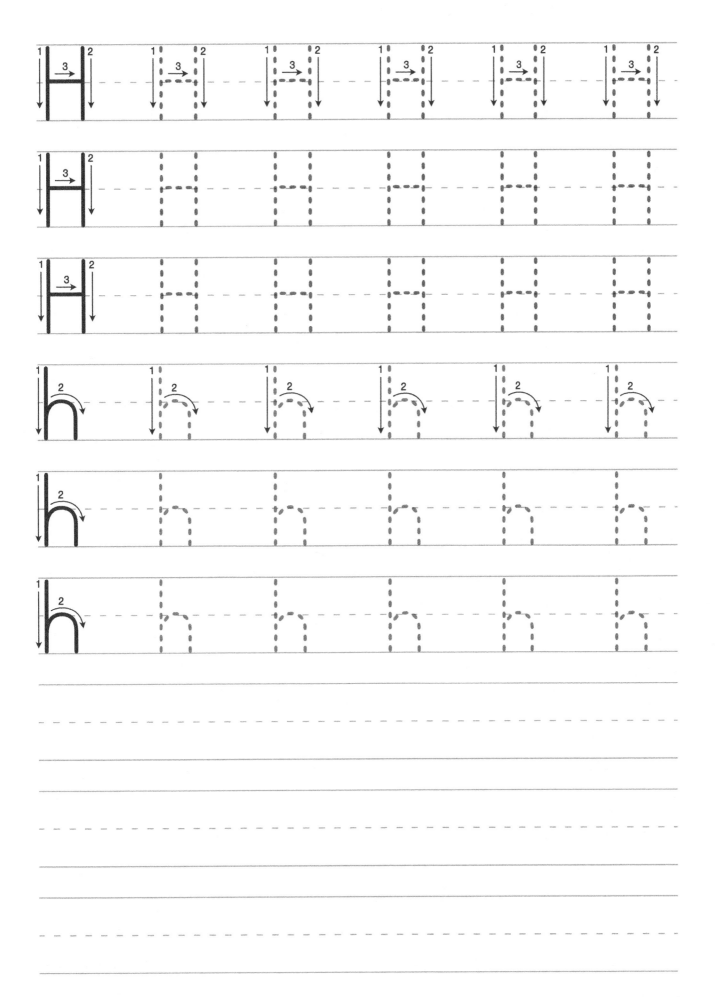

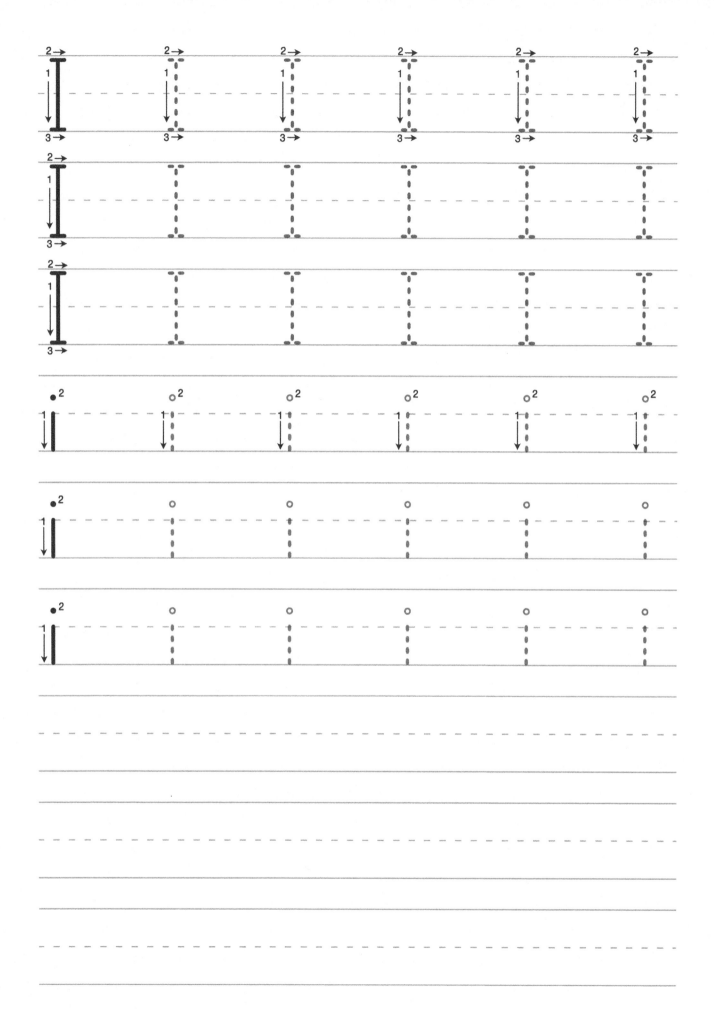

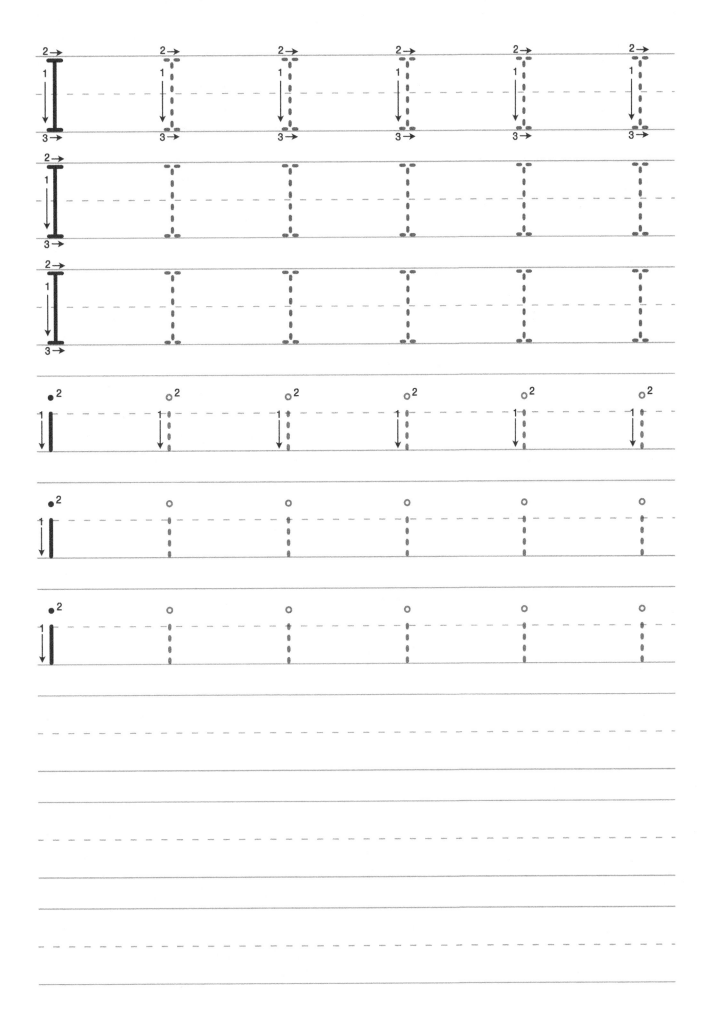

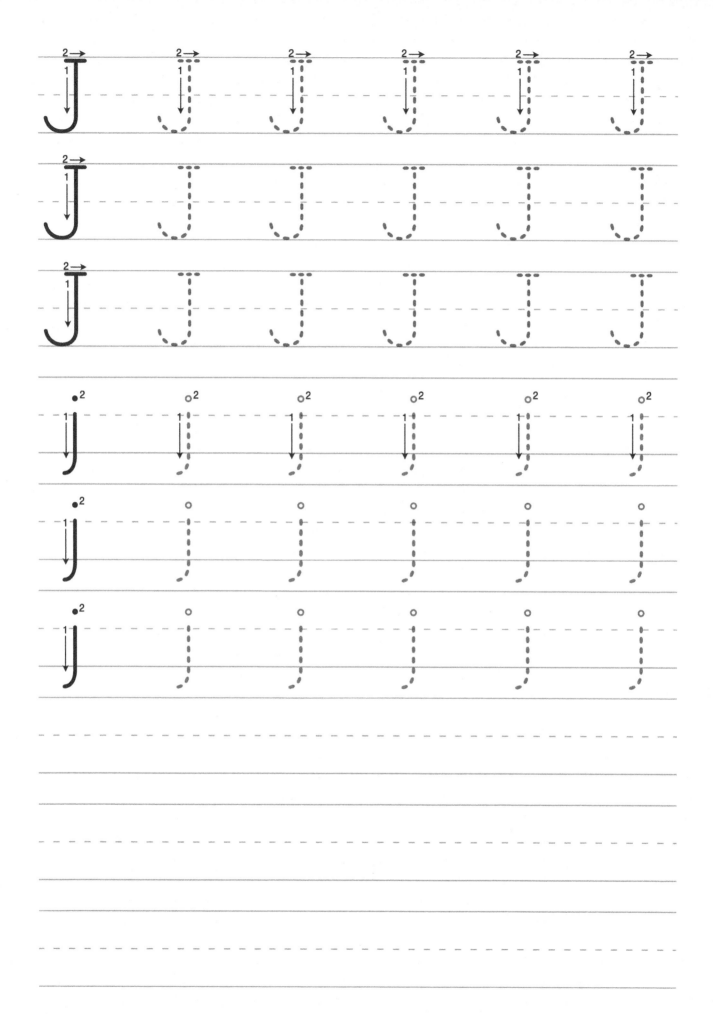

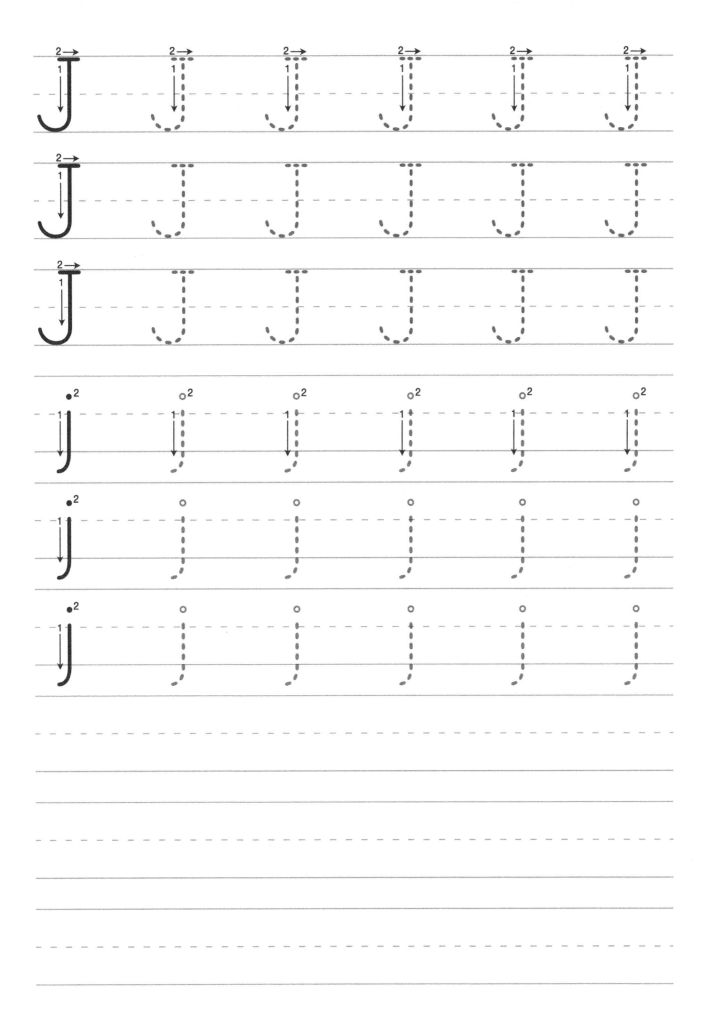

Use the space below to practice your writing.

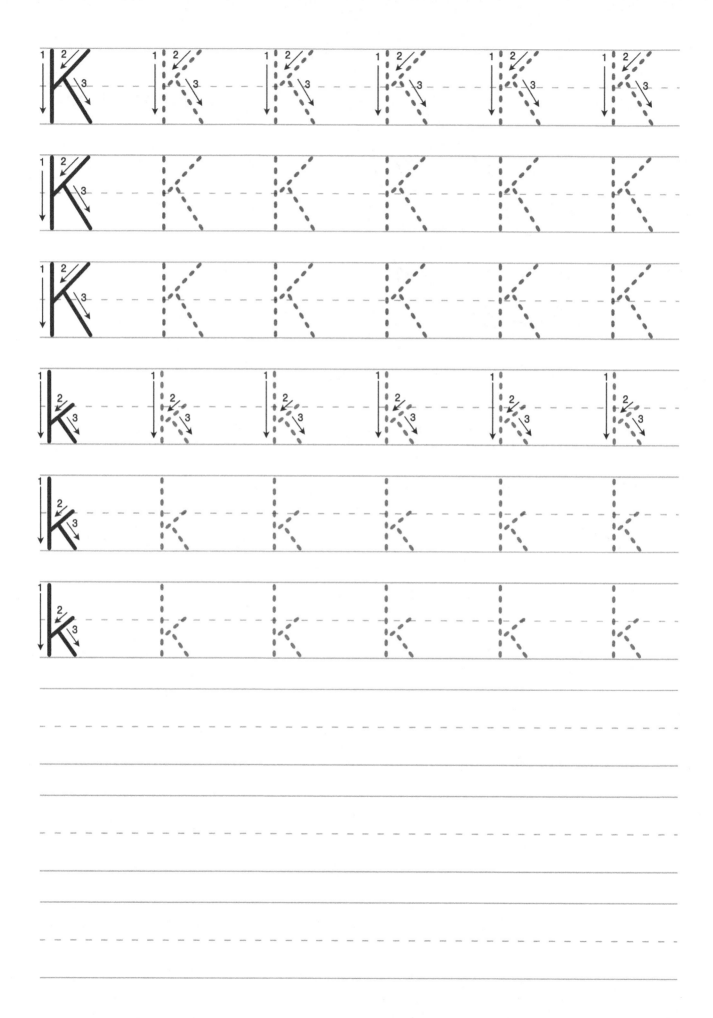

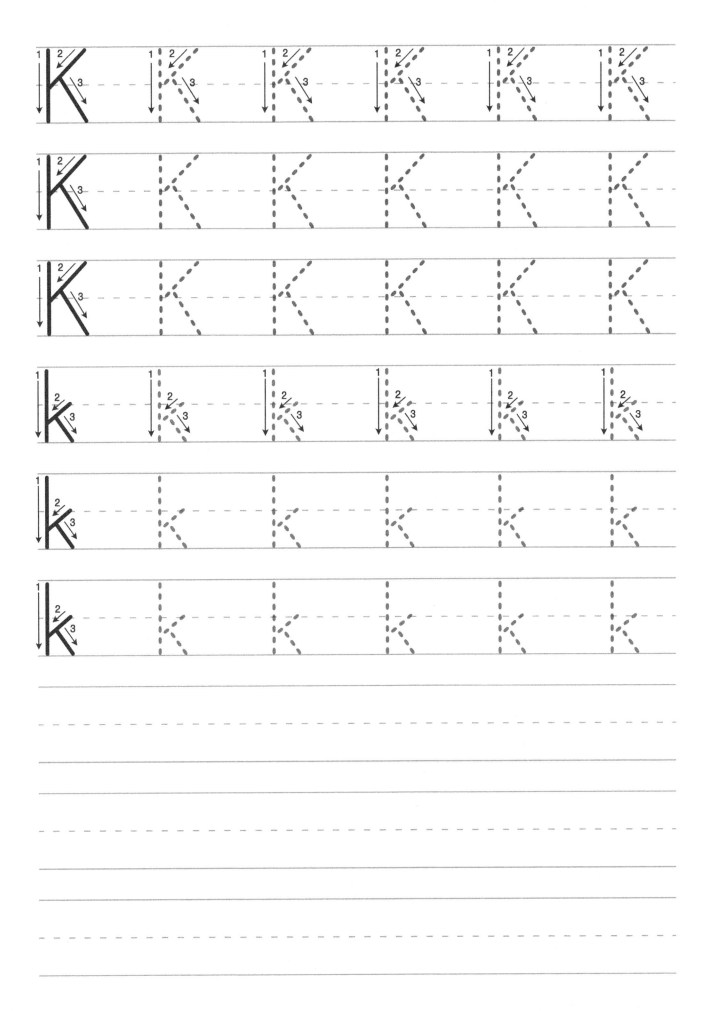

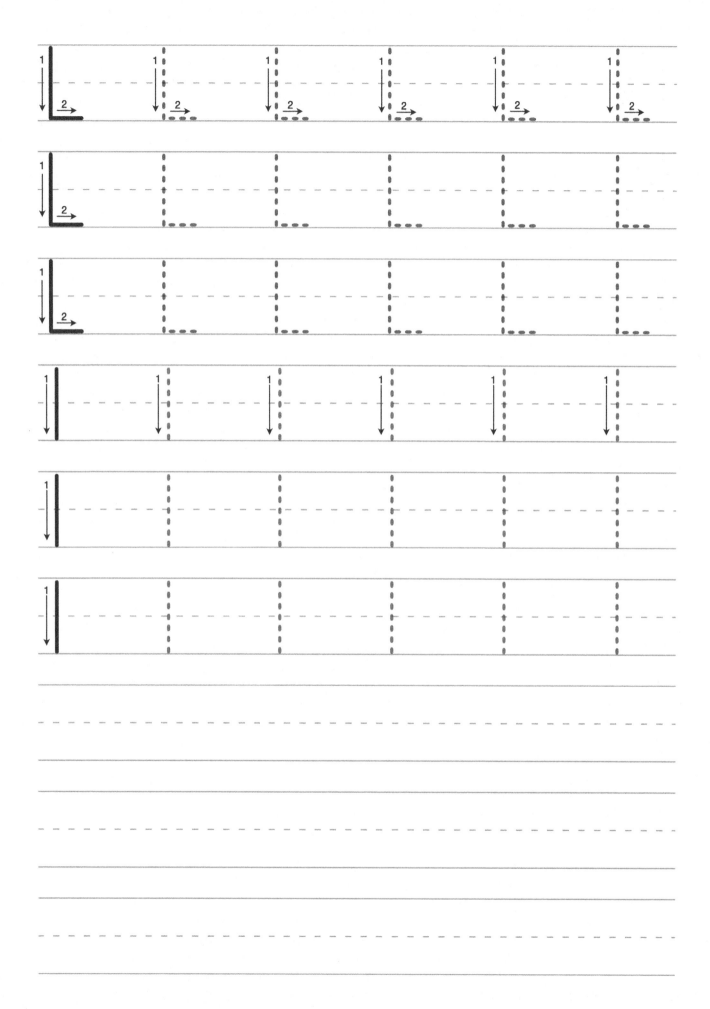

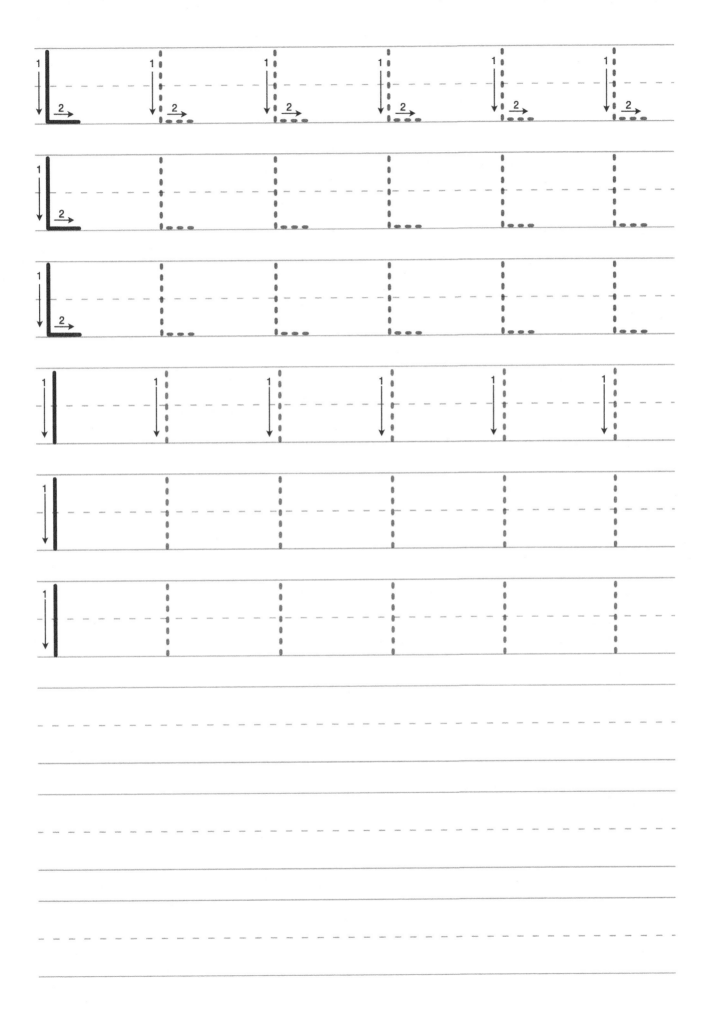

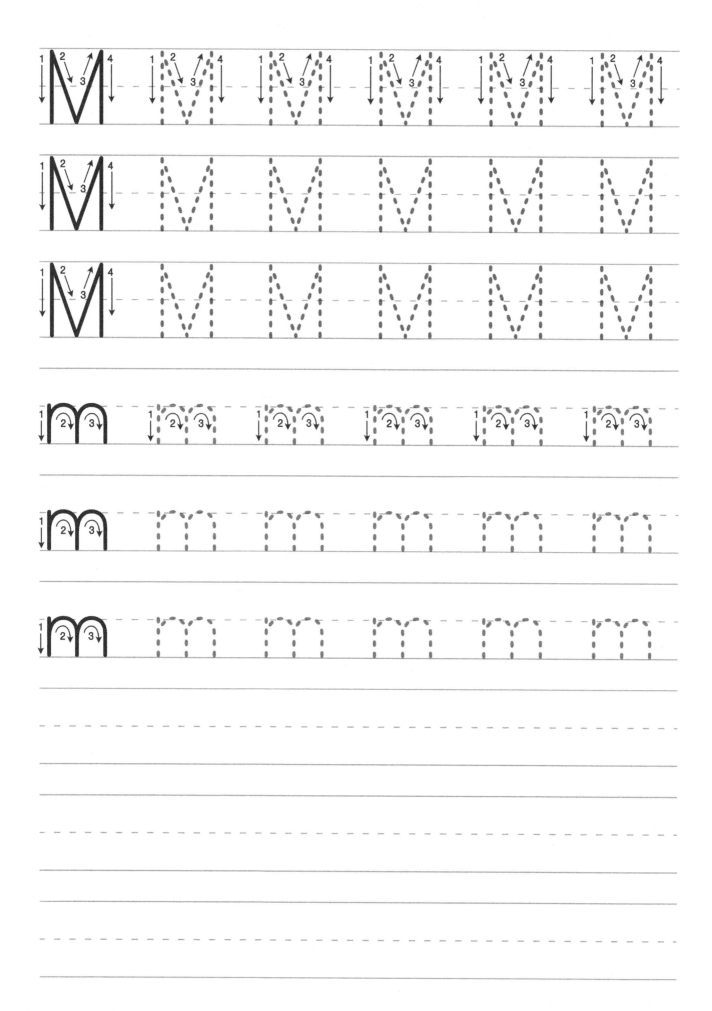

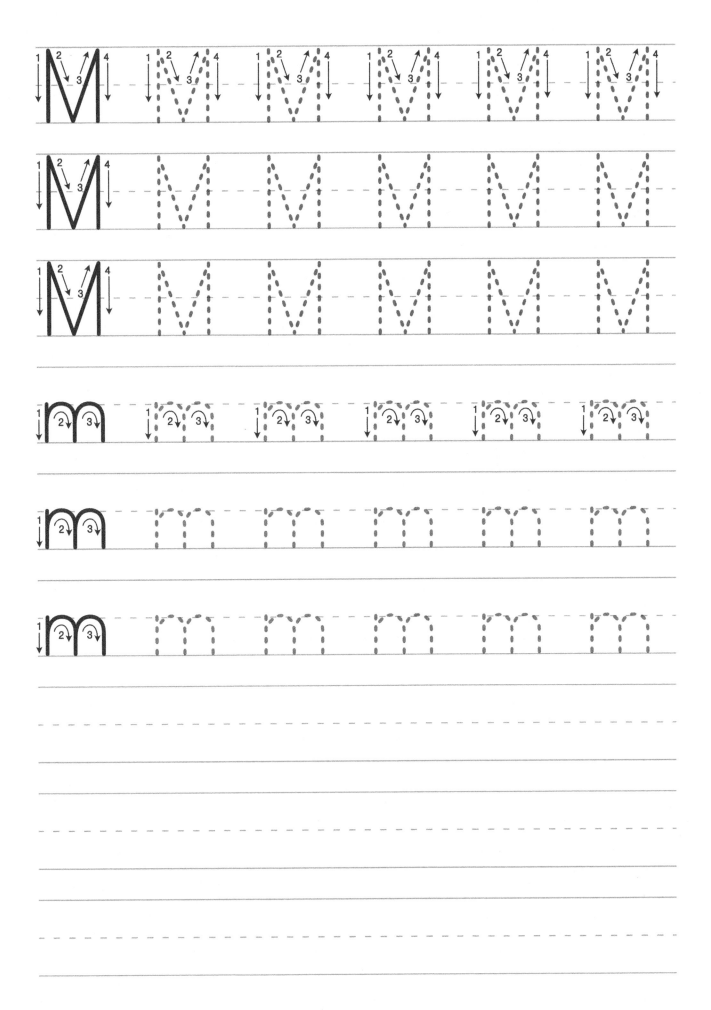

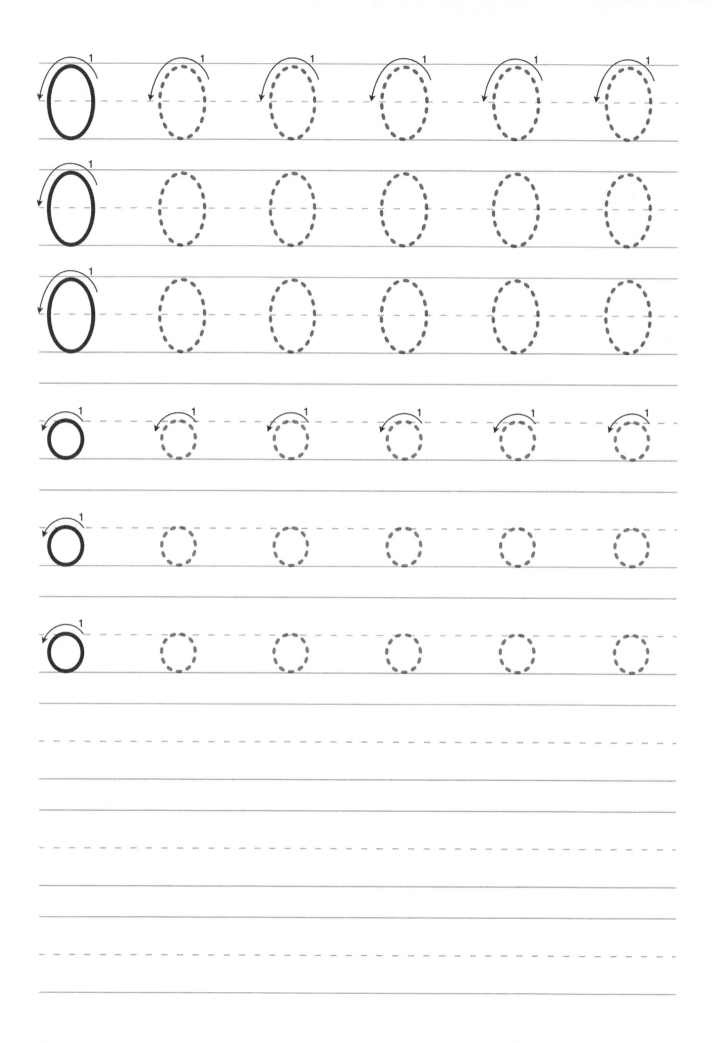

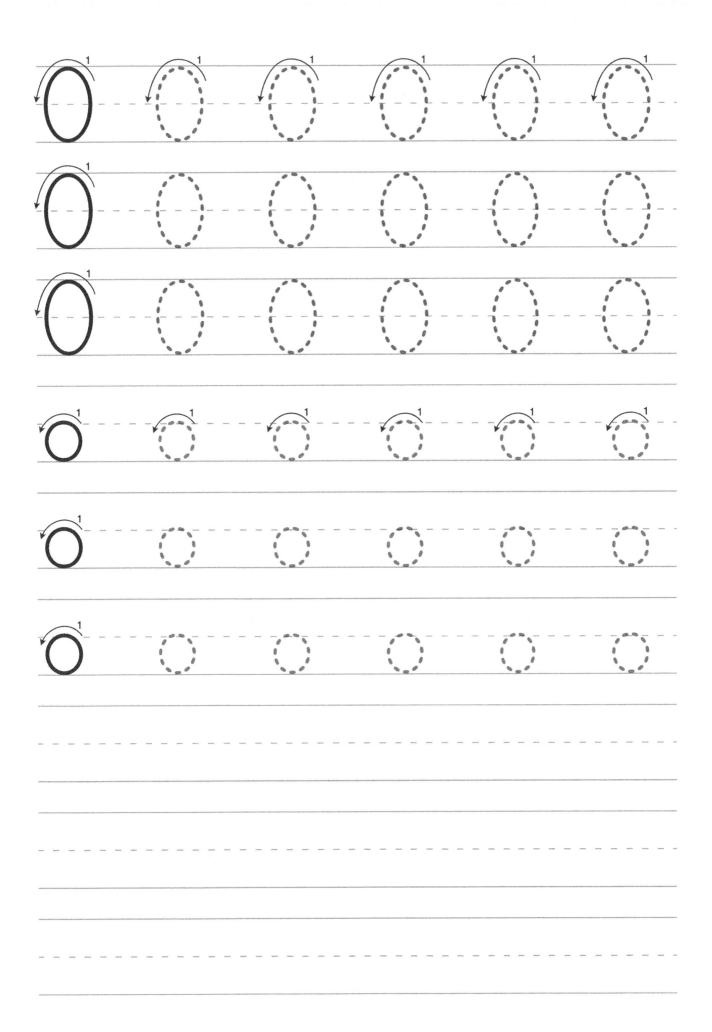

Use the space below to practice your writing.

Color in the Duck

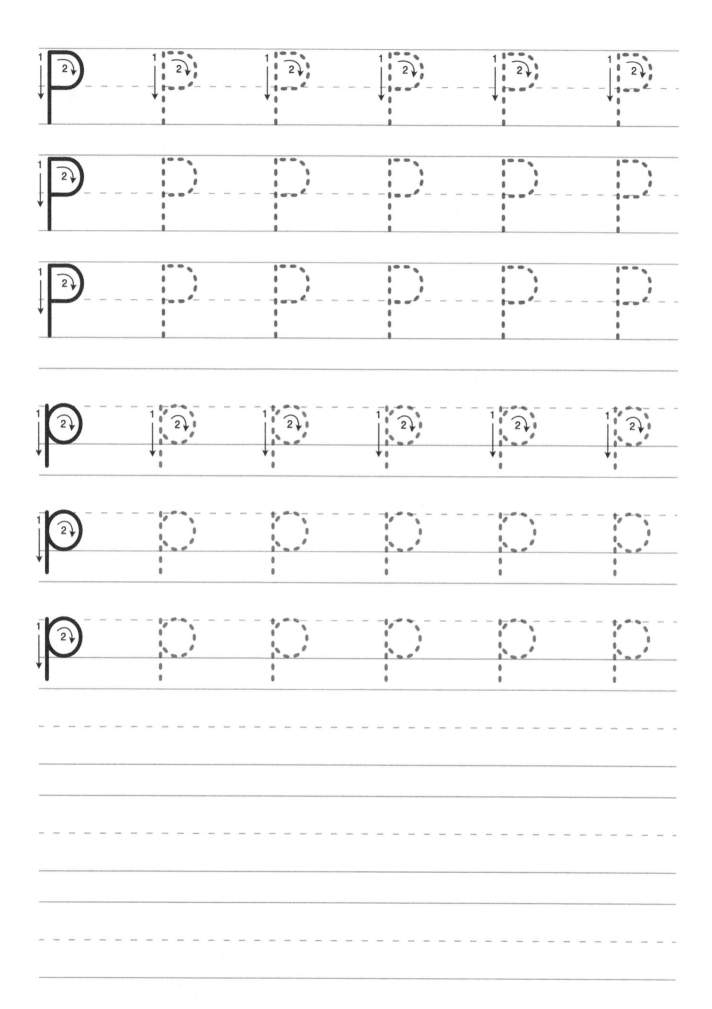

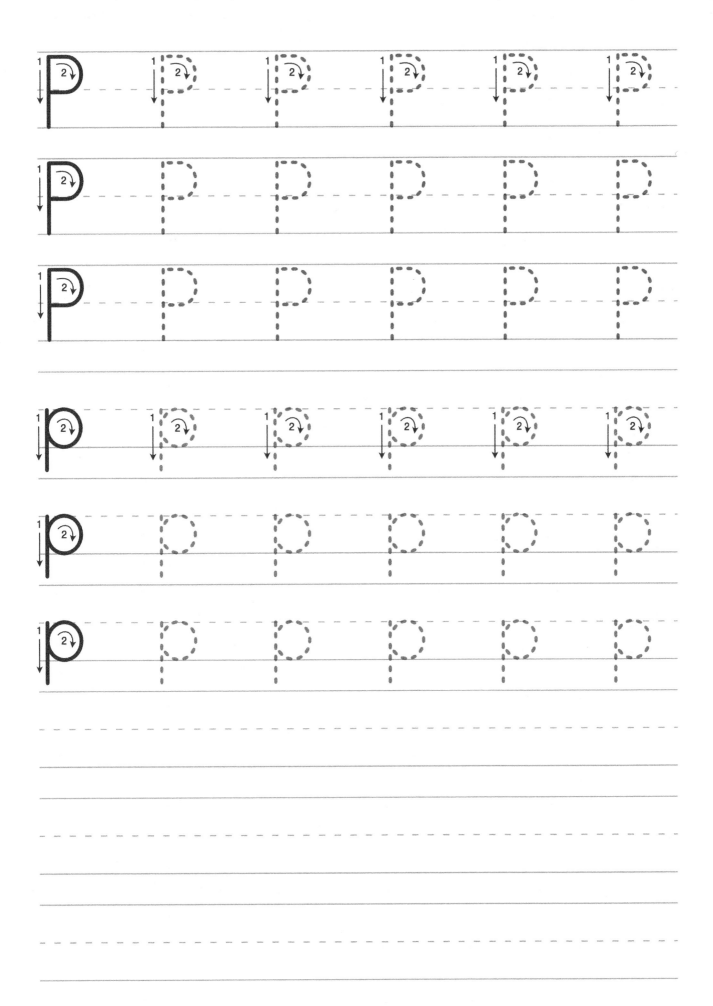

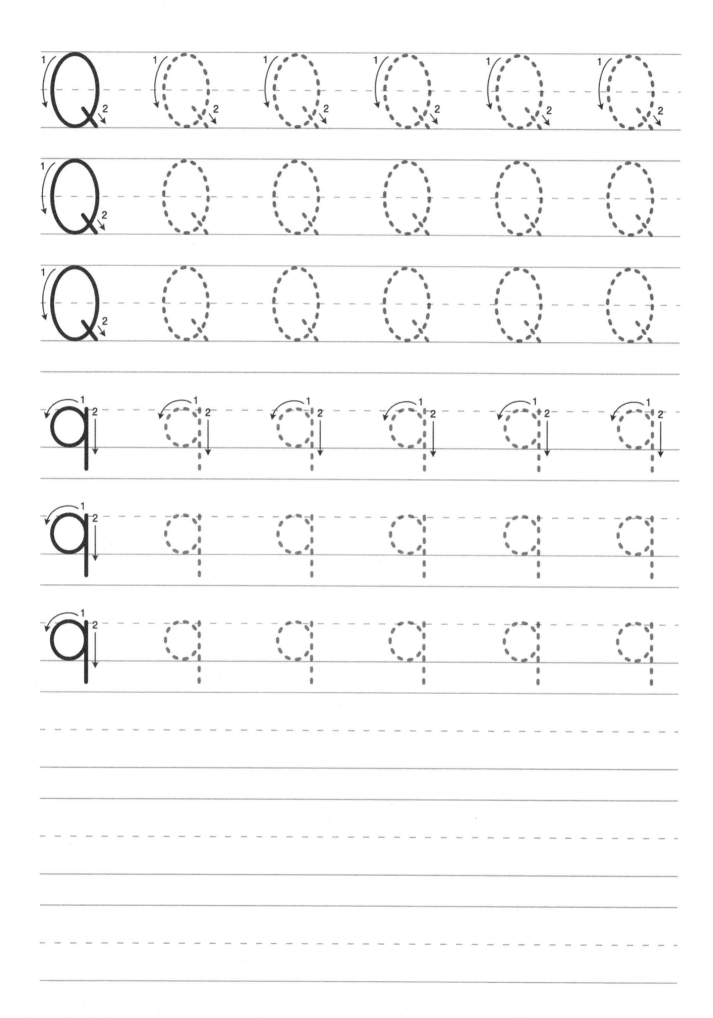

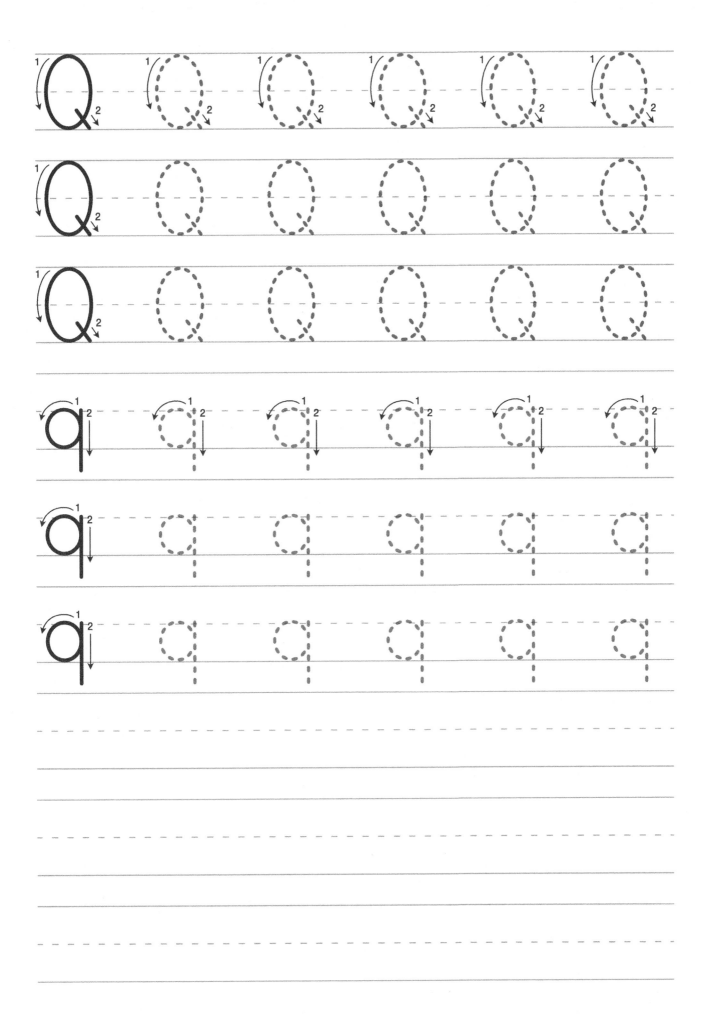

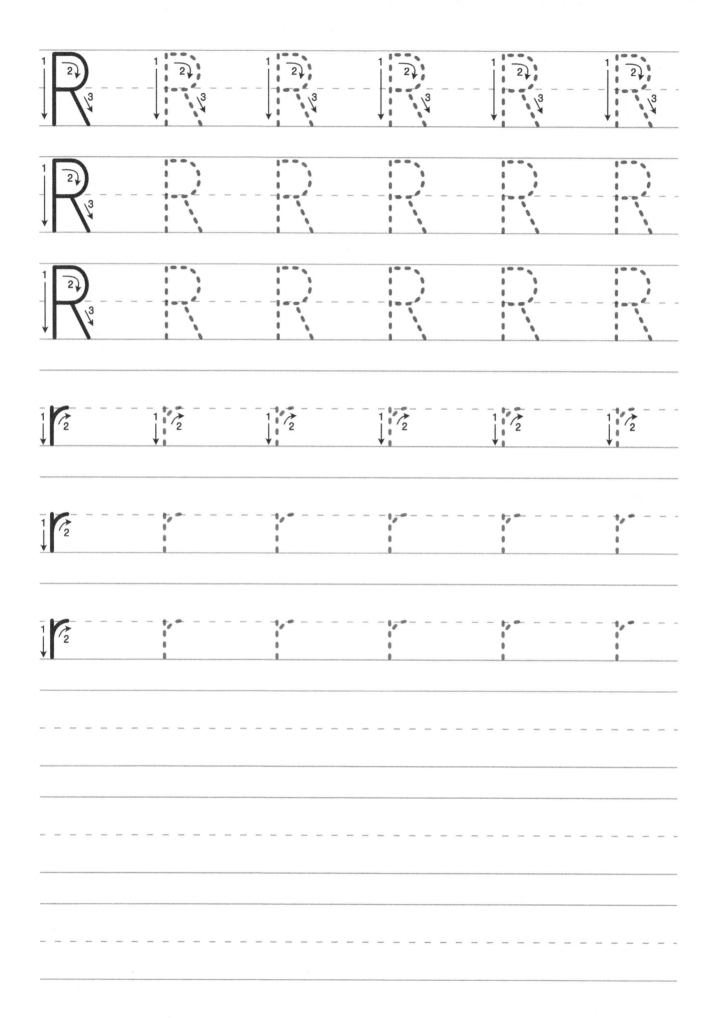

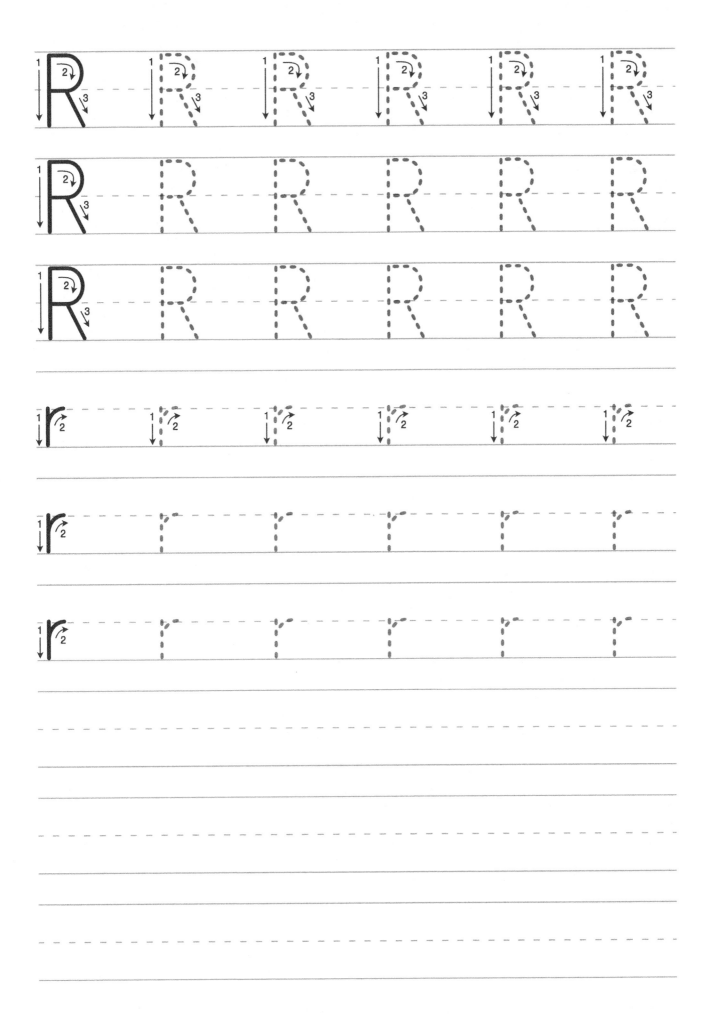

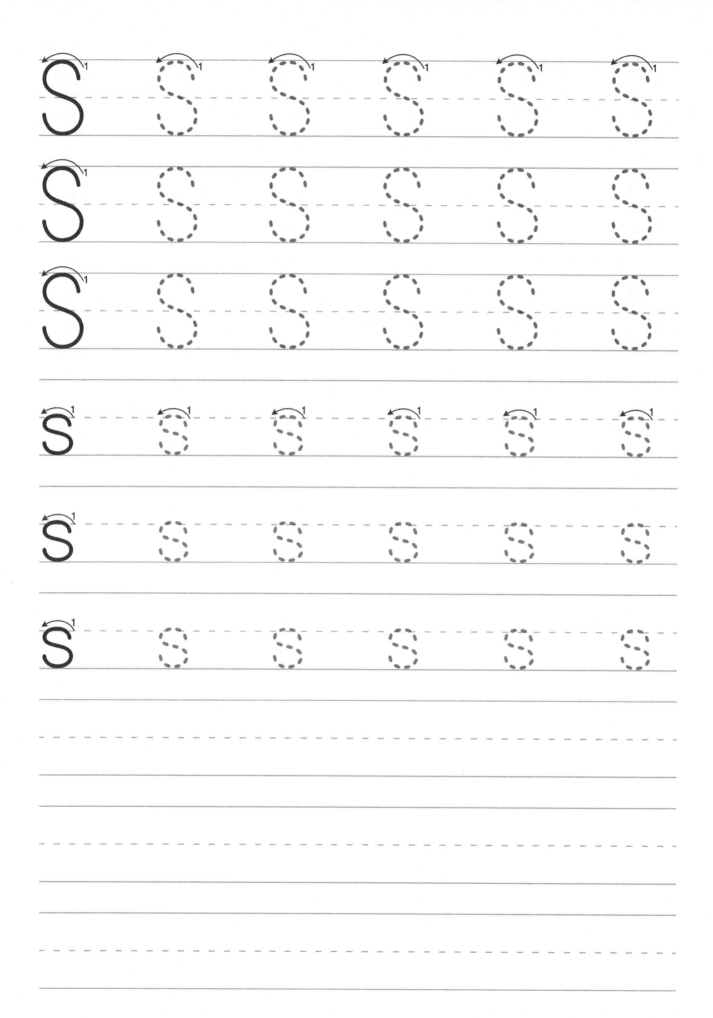

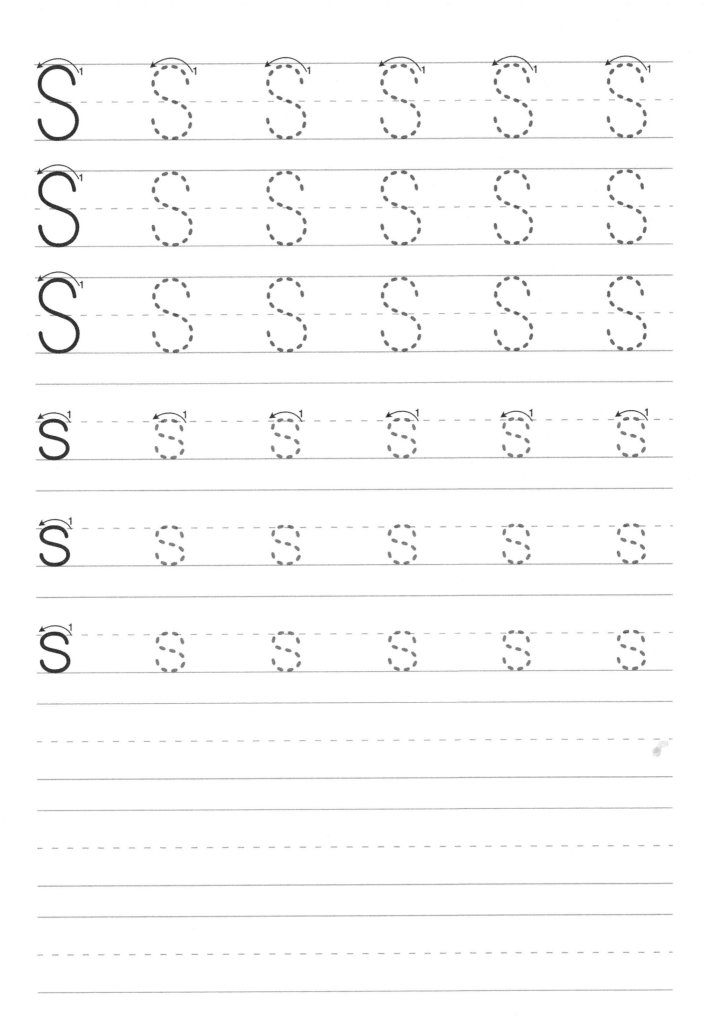

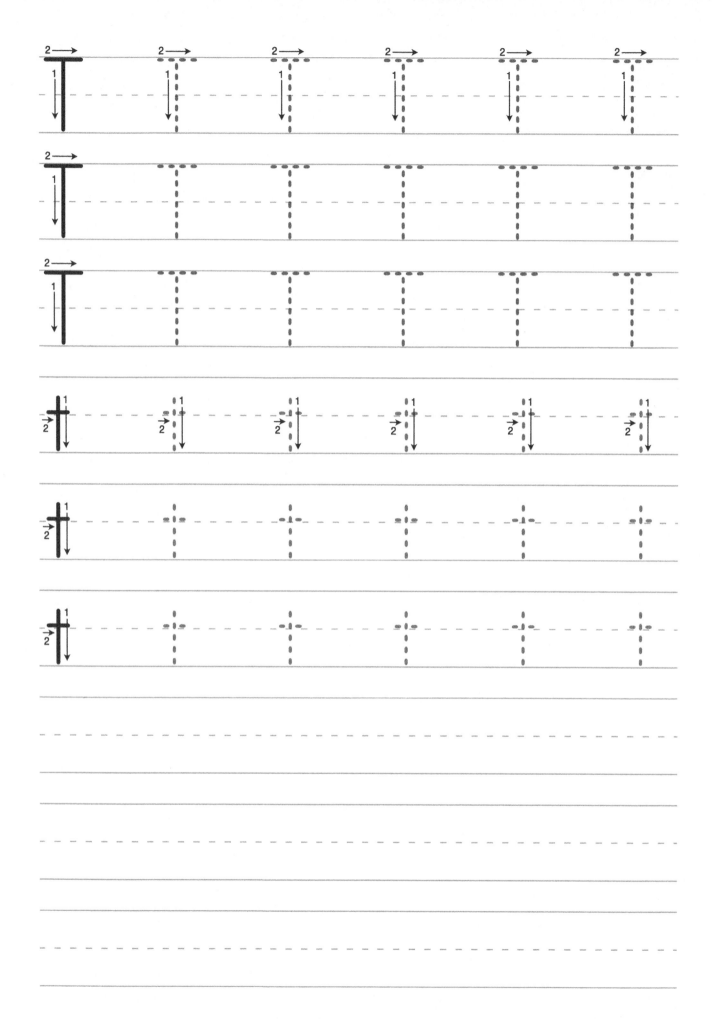

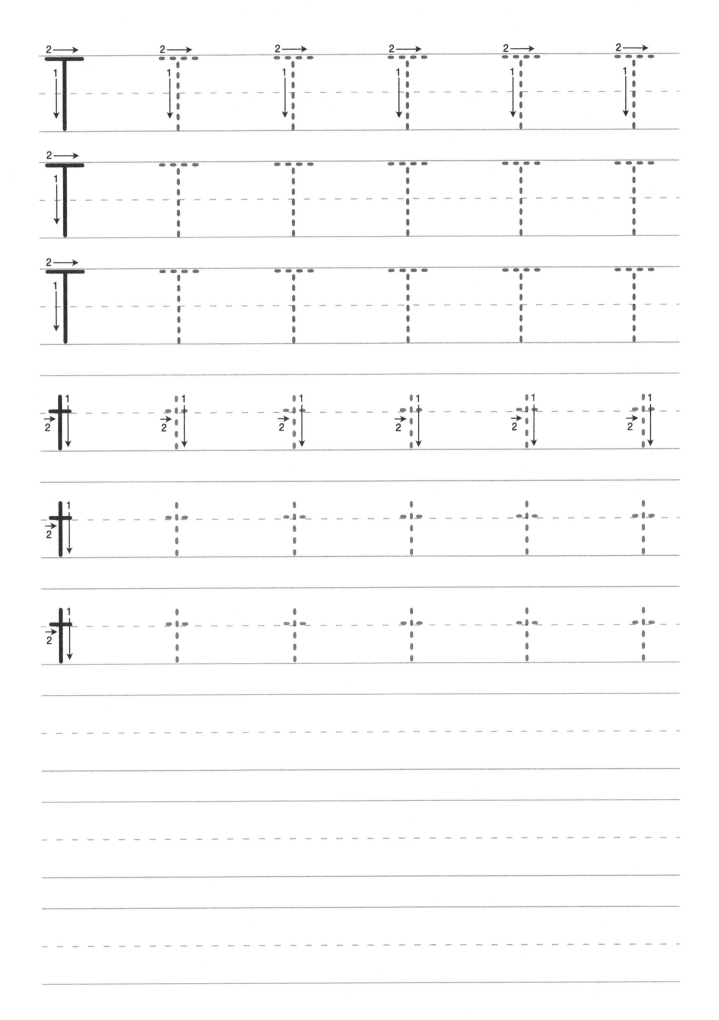

Use the space below to practice your writing.

Color in the Grasshopper

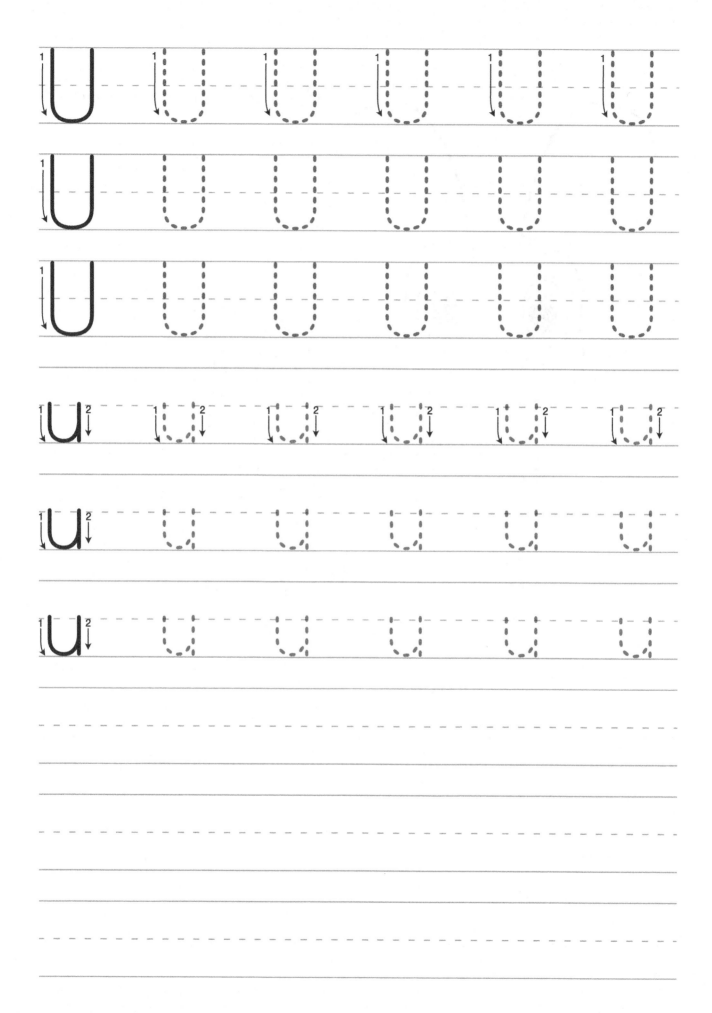

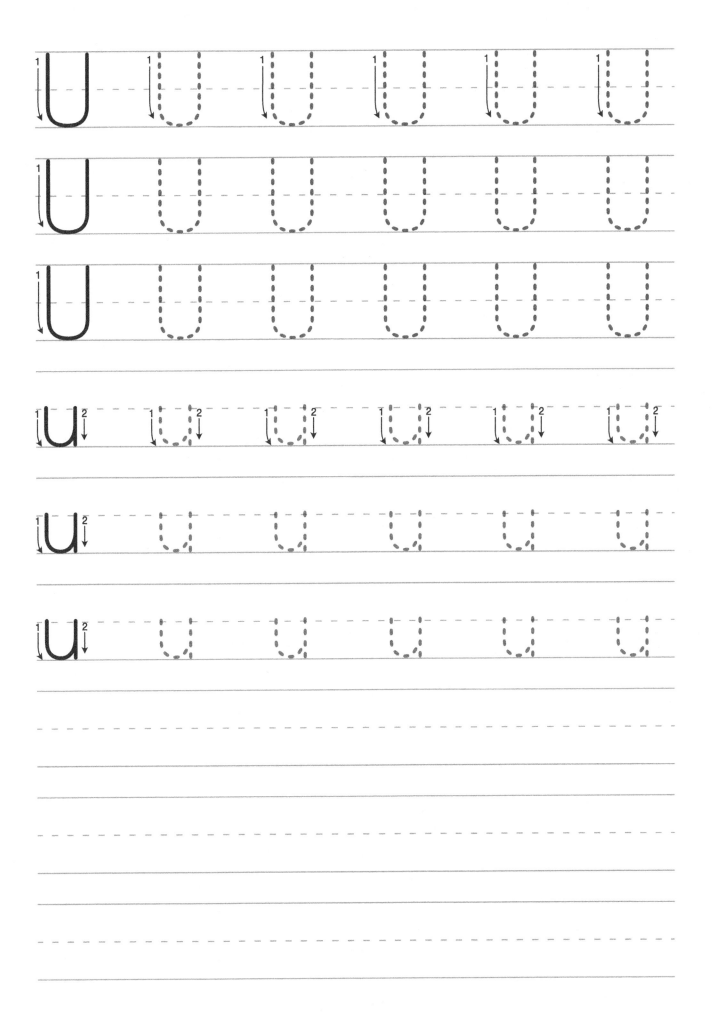

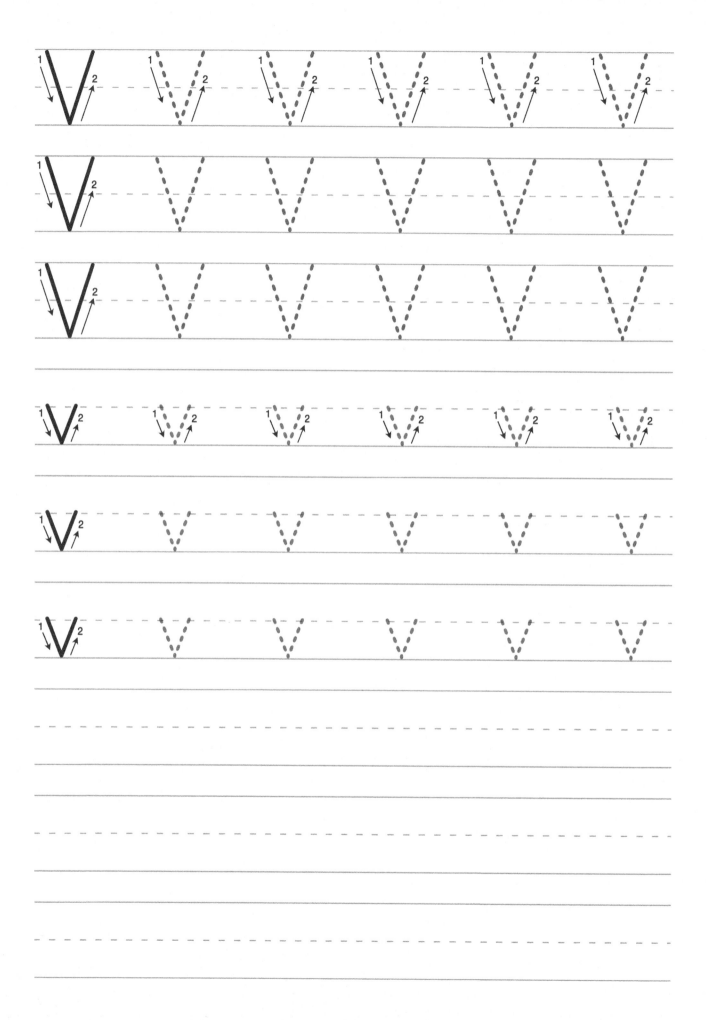

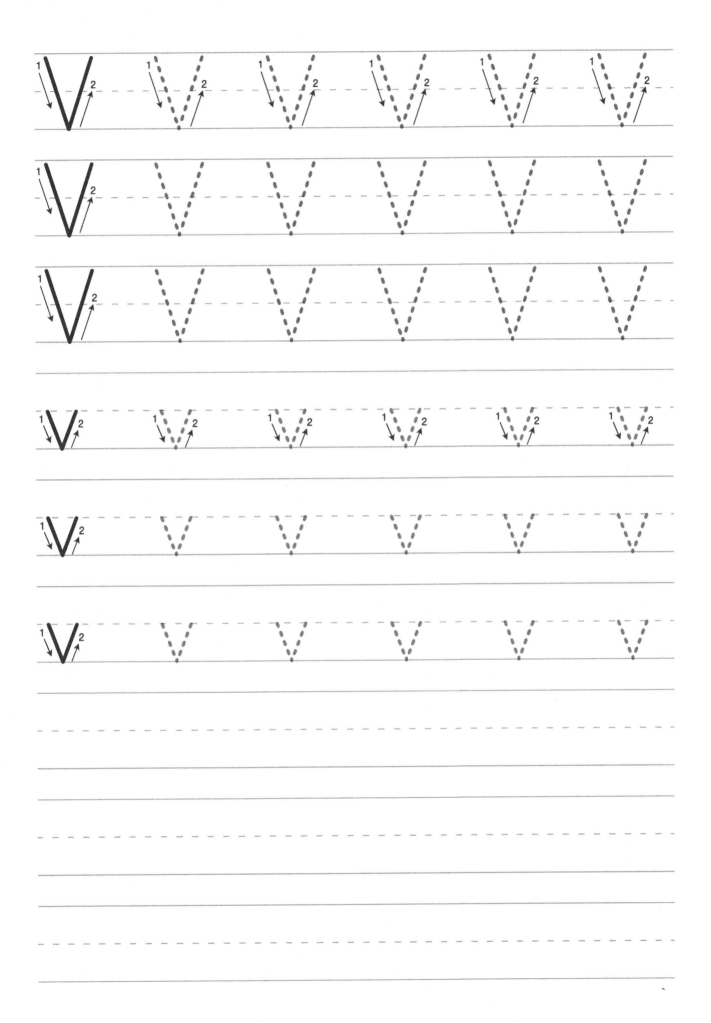

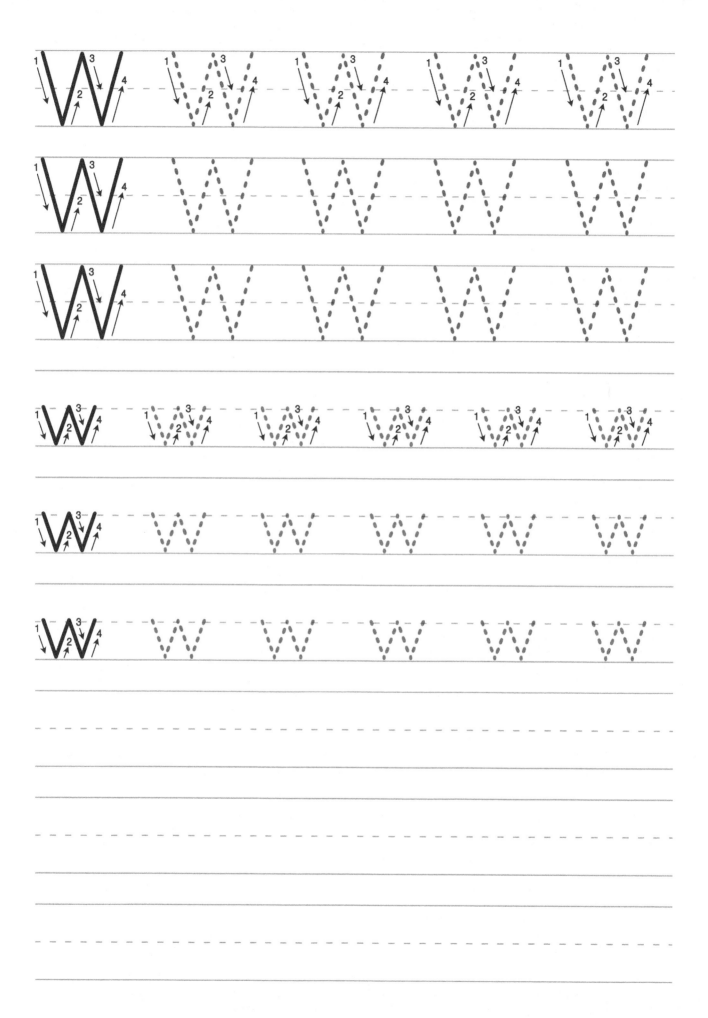

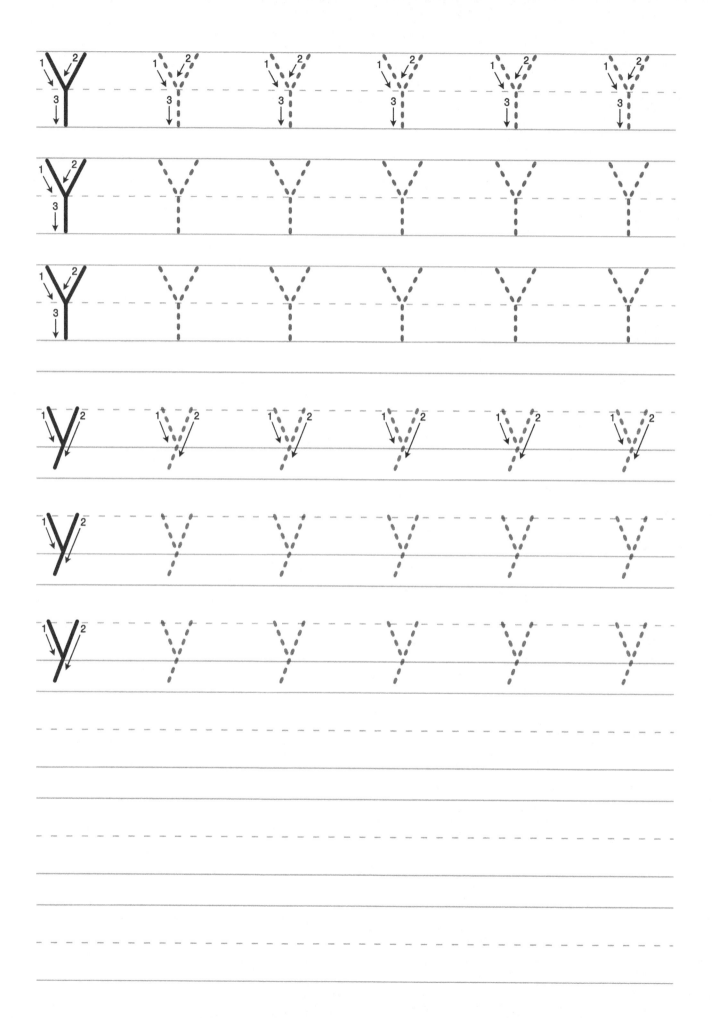

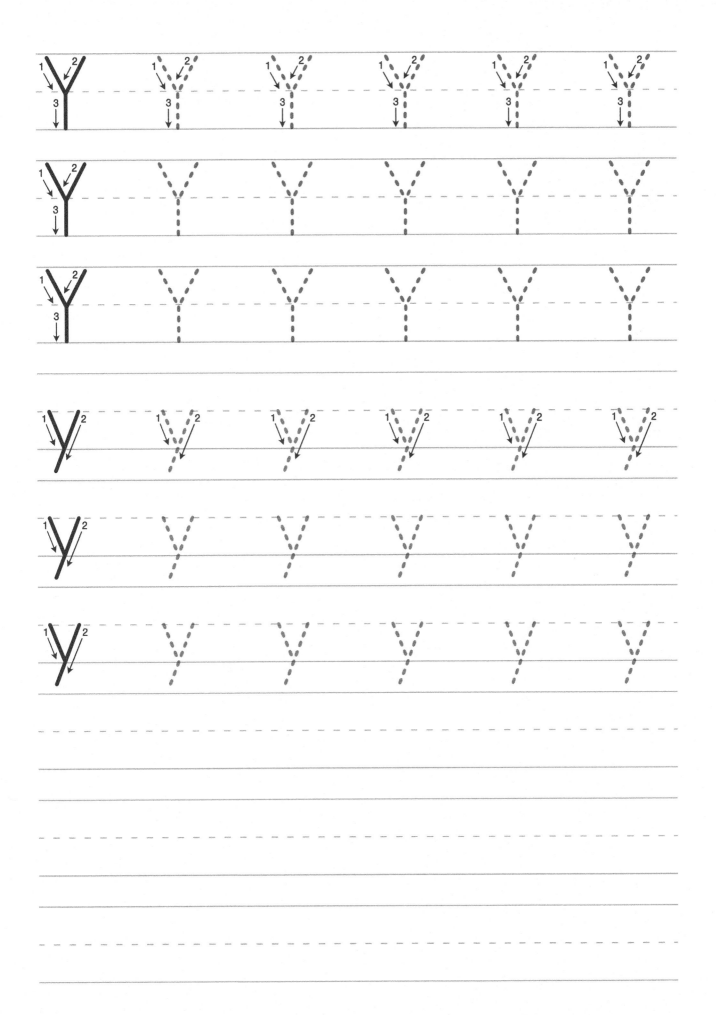

Use the space below to practice your writing.

Read out the alphabet and trace each letter.

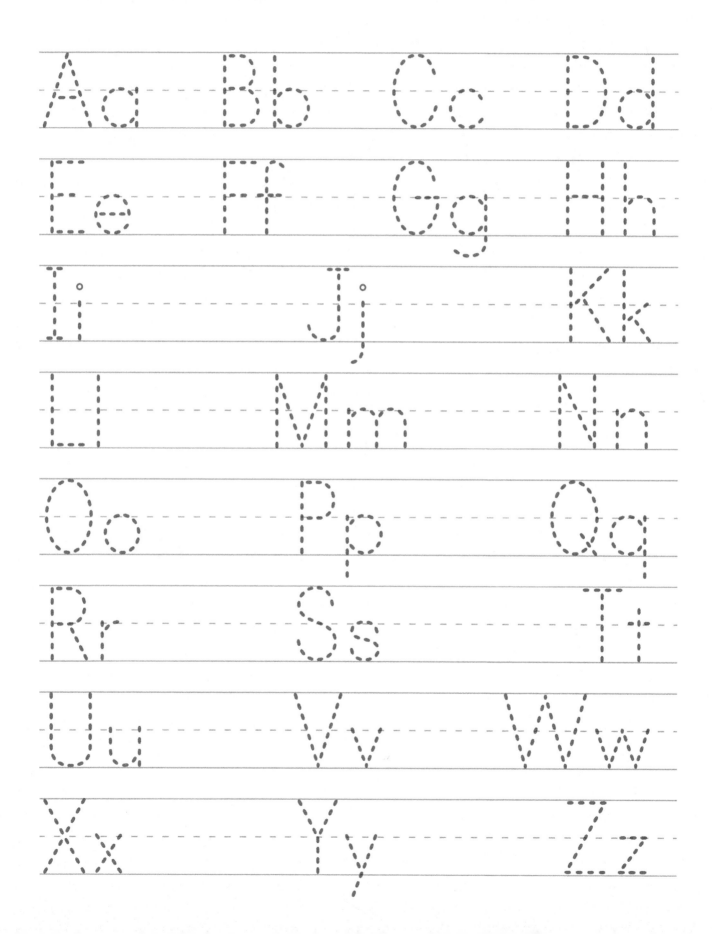

Section 2:
Learning numbers
1-100

We will now learn the numbers 1-100 and practice tracing them and writing them on your own on the blank lines provided.

Color in the Lion

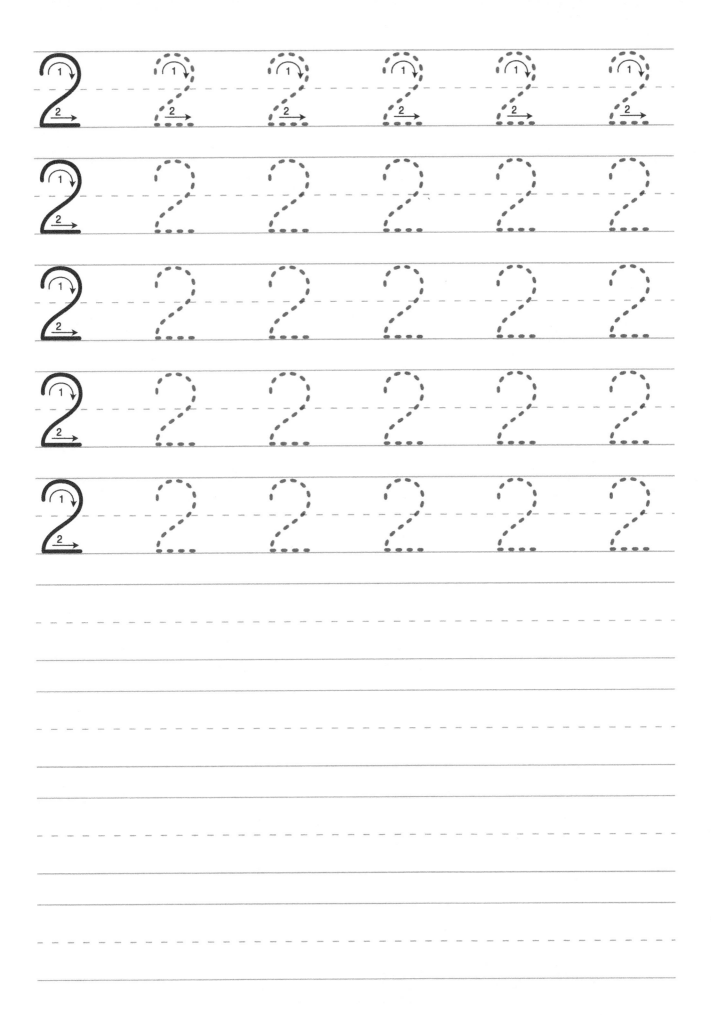

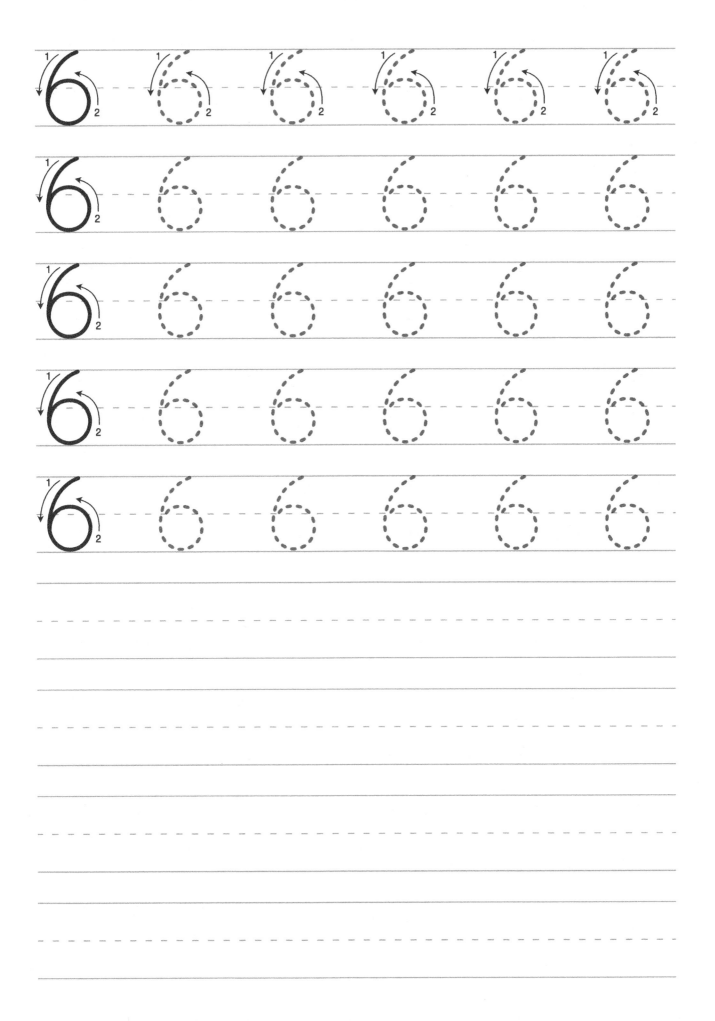

9 9 9 9 9 9

9 9 9 9 9 9

9 9 9 9 9 9

9 9 9 9 9 9

9 9 9 9 9 9

Use the space below to practice your writing.

Color in the Giraffe

11	11	11	11	11
12	12	12	12	12
13	13	13	13	13
14	14	14	14	14
15	15	15	15	15
16	16	16	16	16
17	17	17	17	17
18	18	18	18	18
19	19	19	19	19

20 20 20 20 20

21 21 21 21 21

22 22 22 22 22

23 23 23 23 23

24 24 24 24 24

25 25 25 25 25

26 26 26 26 26

27 27 27 27 27

28 28 28 28 28

29 29 29 29 29

30 30 30 30 30

31 31 31 31 31

32 32 32 32 32

33 33 33 33 33

34 34 34 34 34

35 35 35 35 35

36 36 36 36 36

37 37 37 37 37

38 38 38 38 38

39 39 39 39 39

40 40 40 40 40

41 41 41 41 41

42 42 42 42 42

43 43 43 43 43

44 44 44 44 44

45 45 45 45 45

46 46 46 46 46

47 47 47 47 47

48 48 48 48 48

49 49 49 49 49

50 50 50 50 50

51 51 51 51 51

52 52 52 52 52

53 53 53 53 53

54 54 54 54 54

55 55 55 55 55

56   56   56   56   56

57   57   57   57   57

58   58   58   58   58

59   59   59   59   59

60   60   60   60   60

61   61   61   61   61

62   62   62   62   62

63   63   63   63   63

64   64   64   64   64

65 65 65 65 66

66 66 66 66 66

67 67 67 67 67

68 68 68 68 68

69 69 69 69 69

70 70 70 70 70

71 71 71 71 71

72 72 72 72 72

73 73 73 73 73

74 74 74 74 74

75 75 75 75 75

76 76 76 76 76

77 77 77 77 77

78 78 78 78 78

79 79 79 79 79

80 80 80 80 80

81 81 81 81 81

82 82 82 82 82

83	83 83 83 83
84	84 84 84 84
85	85 85 85 85
86	86 86 86 86
87	87 87 87 87
88	88 88 88 88
89	89 89 89 89
90	90 90 90 90
91	91 91 91 91

92 92 92 92 92

93 93 93 93 93

94 94 94 94 94

95 95 95 95 95

96 96 96 96 96

97 97 97 97 97

98 98 98 98 98

99 99 99 99 99

100 100 100 100

Use the space below to practice your writing.

Use the space below to practice your writing.

Section 3:
Writing and learning a selection of high frequency words

In this section you will learn a small selection of highly used words by tracing and writing them on your own on the blank lines provided.

Color in the Crocodile and Mouse

always   always

and   and   and

are   are   are

ask   ask   ask

better   better

but   but   but

buy   buy   buy

bring   bring   bring

can   can   can

carry   carry   carry

did   did   did

down   down   down

eat   eat   eat

every   every   every

for   for   for

fly   fly   fly

got   got   got

give   give   give

her   her   her

him   him   him

hold   hold   hold

if   if   if

it   it   it

let   let   let

like like like

must must must

make make make

new new new

now now now

number number

our   our   our

only   only   only

put   put   put

pick   pick   pick

run   run   run

read   read   read

saw   saw   saw

show   show   show

the   the   the

tell   tell   tell

use   use   use

upon   upon   upon

very very very

why why why

write write write

you you you

your your your

water water water

Use the space below to practice your writing.

Use the space below to practice your writing.

Section 4:
Writing number words 1-100

Trace the number words and write them on your own on the blank lines provided. Each number word also has the actual number at the beginning of each line so you can associate the number to the number word.

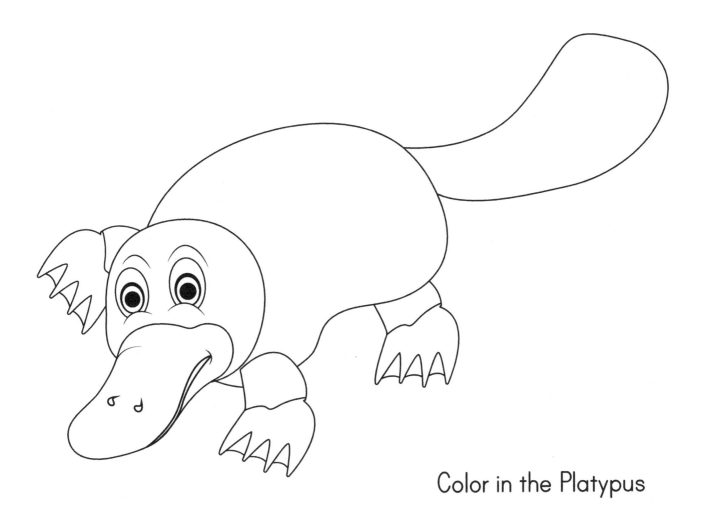

Color in the Platypus

1 one   one

2 two   two

3 three   three

4 four   four

5 five   five

6 six   six

7 seven   seven

8 eight   eight

9 nine   nine

10 ten   ten

11 eleven   eleven

12 twelve   twelve

13 thirteen

thirteen

14 fourteen

fourteen

15 fifteen

fifteen

16 sixteen

sixteen

17 seventeen

seventeen

18 eighteen

eighteen

19 nineteen

nineteen

20 twenty

twenty

21 twenty one

22 twenty two

23 twenty three

24 twenty four

25 twenty five

26 twenty six

27 twenty seven

28 twenty eight

29 twenty nine

30 thirty

31 thirty one

32 thirty two

33 thirty three

34 thirty four

35 thirty five

36 thirty six

37 thirty seven

38 thirty eight

39 thirty nine

40 forty

41 forty one

42 forty two

43 forty three

44 forty four

Use the space below to practice your writing.

45 forty five

46 forty six

47 forty seven

48 forty eight

49 forty nine

50 fifty

51 fifty one

52 fifty two

53 fifty three

54 fifty four

55 fifty five

56 fifty six

57 fifty seven

58 fifty eight

59 fifty nine

60 sixty

61 sixty one

62 sixty two

63 sixty three

64 sixty four

65 sixty five

66 sixty six

67 sixty seven

68 sixty eight

69   sixty nine

70   seventy

71   seventy one

72   seventy two

73   seventy three

74   seventy four

75 seventy five

76 seventy six

77 seventy seven

78 seventy eight

79 seventy nine

80 eighty

81 eighty one

82 eighty two

83 eighty three

84 eighty four

85 eighty five

86 eighty six

87 eighty seven

88 eighty eight

89 eighty nine

90 ninety

91 ninety one

92 ninety two

93 ninety three

94 ninety four

95 ninety five

96 ninety six

97 ninety seven

98 ninety eight

99 ninety nine

100 one hundred

Use the space below to practice your writing.

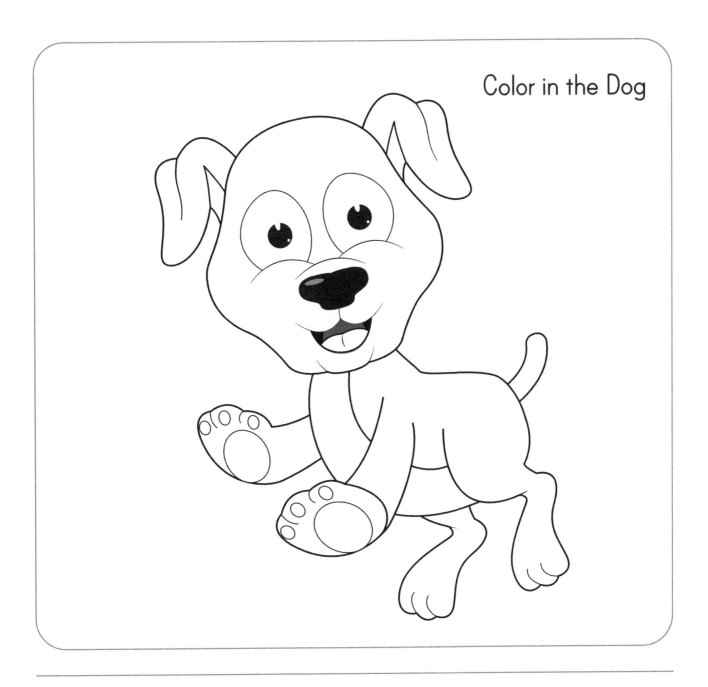

Made in United States
North Haven, CT
04 December 2021